This series fosters information exchange and discussion on all aspects of manufacturing and surface engineering for modern industry. This series focuses on manufacturing with emphasis in machining and forming technologies, including traditional machining (turning, milling, drilling, etc.), non-traditional machining (EDM, USM, LAM, etc.), abrasive machining, hard part machining, high speed machining, high efficiency machining, micromachining, internet-based machining, metal casting, joining, powder metallurgy, extrusion, forging, rolling, drawing, sheet metal forming, microforming, hydroforming, thermoforming, incremental forming, plastics/composites processing, ceramic processing, hybrid processes (thermal, plasma, chemical and electrical energy assisted methods), etc. The manufacturability of all materials will be considered, including metals, polymers, ceramics, composites, biomaterials, nanomaterials, etc. The series covers the full range of surface engineering aspects such as surface metrology, surface integrity, contact mechanics, friction and wear, lubrication and lubricants, coatings an surface treatments, multiscale tribology including biomedical systems and manufacturing processes. Moreover, the series covers the computational methods and optimization techniques applied in manufacturing and surface engineering. Contributions to this book series are welcome on all subjects of manufacturing and surface engineering. Especially welcome are books that pioneer new research directions, raise new questions and new possibilities, or examine old problems from a new angle. To submit a proposal or request further information, please contact Dr. Mayra Castro, Publishing Editor Applied Sciences, via mayra.castro@springer.com or Professor J. Paulo Davim, Book Series Editor, via pdavim@ua.pt

More information about this series at http://www.springer.com/series/10623

Daniel Afonso · Ricardo Alves de Sousa ·
Ricardo Torcato · Liliana Pires

Incremental Forming as a Rapid Tooling Process

Daniel Afonso
TEMA: Centre for Mechanical Technology
and Automation, School of Design,
Management and Production
Technologies Northern Aveiro
University of Aveiro
Oliveira de Azeméis, Portugal

Ricardo Torcato
CICECO: Aveiro Institute of Materials,
School of Design, Management and
Production Technologies Northern Aveiro
University of Aveiro
Oliveira de Azeméis, Portugal

Ricardo Alves de Sousa
TEMA: Centre for Mechanical Technology
and Automation, Department of Mechanical
Engineering
University of Aveiro
Aveiro, Portugal

Liliana Pires
CICECO: Aveiro Institute of Materials,
School of Design, Management and
Production Technologies Northern Aveiro
University of Aveiro
Oliveira de Azeméis, Portugal

ISSN 2191-530X   ISSN 2191-5318 (electronic)
SpringerBriefs in Applied Sciences and Technology
ISSN 2365-8223   ISSN 2365-8231 (electronic)
Manufacturing and Surface Engineering
ISBN 978-3-030-15359-5   ISBN 978-3-030-15360-1   (eBook)
https://doi.org/10.1007/978-3-030-15360-1

Library of Congress Control Number: 2019934526

This Springer imprint is published by the registered company Springer Nature Switzerland AG
The registered company address is: Gewerbestrasse 11, 6330 Cham, Switzerland

Preface

The rapid tooling (RT) notion has been discussed since the beginning of the additive manufacturing processes, formerly designated as rapid prototyping. The concept consists of using fast smart manufacturing technologies to develop tools to process other materials. This novelty allows decreasing the time to market, decrease tooling cost and increase tooling complexity and consequential part design freedom.

Mainly due to their novelty and the technologies used in most processes, RT has been majorly associated with additive manufacturing (AM), commonly called as fast free-form fabrication. Nevertheless, incremental sheet forming (ISF) processes can be seen as rapid prototyping processes, and so also considered rapid manufacturing methodologies or fast free-form fabrication processes. Besides, being compatible with flexible manufacturing systems, with parts manufactured from computer-aided design (CAD) models without considerable dedicated tools in a short time, ISF processes can be seen analogously to AM technologies. These processes encounter industrial applications not only in prototyping or part manufacturing but also in tool development and fabrication. Thus, it is reasonable to apply the RT term when describing the fabrication of tools for different industrial processes using ISF techniques.

Aveiro, Portugal Daniel Afonso
January 2019

Contents

Acronyms

3DP	Three-Dimensional Printing or 3D Printing
AISF	Asymmetric Incremental Sheet Forming
AM	Additive Manufacturing
CAD	Computer-Aided Design
CAE	Computer-Aided Engineering
CAM	Computer-Aided Manufacturing
CNC	Computer Numerical Control
DMLS	Direct Metal Laser Sintering
DMP	Direct Metal Printing
DPIF	Dual-Point Incremental Forming
EDM	Electrical Discharge Machining
FFF	Fused Filament Fabrication
FFL	Forming Fracture Line
FLC	Forming Limit Curve
FLD	Forming Limit Diagram
ISF	Incremental Sheet Forming
MUD	Master Unit Die
PEL	Profiled Edge Laminae
RIM	Reaction Injection Moulding
RT	Rapid Tooling
RTV	Room Temperature Vulcanization
SL	Stereolithography
SLM	Selective Laser Melting
SLS	Selective Laser Sintering
SPIF	Single-Point Incremental Forming
TPIF	Two-Point Incremental Forming
UAM	Ultrasonic Additive Manufacturing
UV	Ultraviolet
VARTM	Vacuum-Assisted Resin Transfer Moulding

Chapter 1
Fundamentals of Rapid Tooling

1.1 Concept and Classifications

Due to the pressure of highly competitive markets, the industry is driven to compete effectively by reducing manufacturing times and costs while assuring high-quality products and services. Besides, environmentally responsible goals also affect decisions on industrial manufacturing systems. It is now generally accepted to have rapid changes in product models and production volume, which calls for a redefinition of the product design and development techniques, as well as changes in the conventional manufacturing processes [1].

Product development takes advantage of the use of CAD systems to define the geometry and its various dimensional characteristics. Besides, the product's feasibility can be predicted using CAE software for the analysis of product performance and for the simulation of manufacturing processes without the need for physical prototypes. While these iterations strongly improve the probability of success, in many cases, a physical assessment of the real component is still needed. This often requires the creation of prototypes and tools to be produced, becoming one of the most time-consuming and costly phases in the development of new products [1–3].

During the last decades, several new smart manufacturing processes have been developed, with great potential for the fabrication of unique parts, commonly named rapid manufacturing methodologies. These processes include improvements on traditional manufacturing technologies such as in computer numerical control (CNC) milling, and the emergence of new technologies like the additive manufacturing (AM) systems and the incremental sheet forming (ISF) processes. These systems find their applicability not only in the development of prototypes or small volume production but also for tooling fabrication. Thus, new applications are referred to as rapid tooling (RT) techniques and aim to reduce time to market and increase the competitive edge [4]. The leading characteristics of a RT process should ensure [5]:

- Tooling time is much shorter than for a conventional tool. Typically, time to first articles is below one-fifth that of the conventional tooling.
- Tooling cost is much less than for a conventional tool. Cost can be below 5 percent of conventional tooling cost.

© The Author(s), under exclusive licence to Springer Nature Switzerland AG 2019
D. Afonso et al., *Incremental Forming as a Rapid
Tooling Process*, Manufacturing and Surface Engineering,
https://doi.org/10.1007/978-3-030-15360-1_1

Fig. 1.1 Rapid tooling
classifications

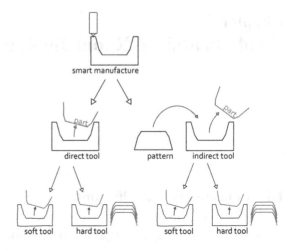

- Tool life is considerably less than for a conventional tool.
- Tolerances are coarser than in the conventional tools and have worst surface finishing.

RT finds its applicability for the development of prototype parts, for small batches production or for first product fabrication before the development of full production tools. By definition, a true prototype is an object produced in the intended material, by the final method of production. The use of RT allows the development of true prototypes, otherwise hard or impossible to obtain. For production scenarios, it is not reasonable to develop conventional tooling for small volume production. Thus, low-cost RT fills this gap between unique products and large volume mass production. Finally, the development of production tools is typically a time-consuming process that can delay a product launch. In these cases, RT can assume an important role in the manufacturing of the first units, assuming the concept of bridge tooling [2].

Two main classifications of RT are indirect and direct methods. Indirect tooling processes use patterns to produce tools. Direct tooling implies the ability to fabricate a tool directly from a rapid manufacturing machine. Besides, the tooling process can also be classified according to the used materials. If the tooling material can only be used to produce a few part copies before it wears, such process is referred as soft tooling. Hard tooling, on the other hand, involves the production of tools capable of producing thousands of parts. Figure 1.1 presents the four possible classifications for RT applications [1, 4].

A large variety of technologies and techniques can be used for RT development. As mentioned, RT in mainly associated with AM process. Given the diversity of available processes, tool development takes advantage of the fabrication of parts in different materials, including metals, polymers, and ceramics. The inclusion of other manufacturing processes in RT application either in stand-alone processes or in hybrid rapid manufacturing increases, even more, the supply of possible solutions [6, 7].

RT is the term used for the production of tools or tooling components with reduced lead time, as compared to conventional manufacturing techniques, through direct or indirect processes. The direct processes consist in using one technology capable to reproduce tools with similar properties to the actual tools. The indirect processes involve the use of more than one technology, first to produce a pattern and second to obtain the actual tools through the pattern [2, 3, 8].

Conventional manufacturing technologies, such as investment casting and injection moulding, are often used in the production of products. These technologies require tools or tooling components (e.g. moulds and inserts) that need to be developed, specifically for a particular purpose or new product. Machining technologies and heat treatments, frequently used, augment time and costs in the product development process. Sometimes, these factors invalidate the low volume productions and limit the introduction, in a market, of custom solutions or niche products [2].

The adoption of rapid manufacturing technologies allows creating skills on the production of tools for low volume products production, as wells as reduces production costs and time to get products to market. These technologies are applied also to test and validate projects with the manufacturing of 3D physical models. This step allows catching errors early in the product development process and designing better products. Improvements or tool modifications are sometimes required during the manufacturing process. The rapid manufacturing tooling can bring benefits also by producing faster solutions. Additive and subtractive methods can be applied to fabricate tools. The accuracy of the tooling is determined by the type of machines and processes selected and controlled by CAD/computer-aided manufacturing (CAM) digital techniques [1, 2, 9].

A first well-known state-of-the-art paper on RT has been published in 2000 [9]. This report presents the RT principle, based on the use of different AM processes, and presents examples of applications for casting, thermoset processing and thermoplastics processing. An extended review on RT has been published annually since 2006 by Wohlers Associates [10].

Figure 1.2 presents the most relevant RT processes, grouped by technology and materials. While in conventional tooling, the process includes a sampling step where the moulded part is evaluated, in which many of the RT solutions do not offer the same possibility. The tool is fabricated in one shot and modifications may require starting over. Thus, the selection of a RT process finds its bigger significance not only in suiting the processing material and process requirements but also in forecasting the moulding results. According to the method and technology used for the tool development, soft tooling, hard tooling and bridge tooling can be produced. Room temperature vulcanization silicone (RTV) and stereolithography (SL) are indirect and direct technologies mostly used to produce soft tooling, made by soft materials for low production, up to 50 units. For high production volume, hard tooling must be made with durable material. Direct methods such as selective laser sintering (SLS)

Fig. 1.2 Rapid tooling
processes

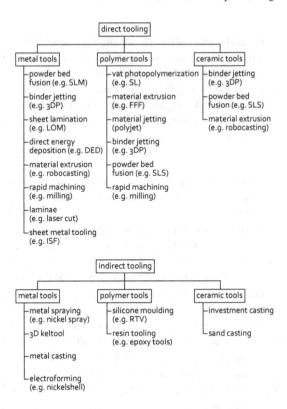

and direct metal printing (DMP) of powder metal materials or incremental sheet forming (ISF) of sheet metal are used to produce these type of tooling [1, 11].

1.2 Direct AM Rapid Tooling Processes

AM is applied in RT to manufacture tools or tooling components with reduced lead times. AM is the standard term used for manufacturing technologies which produce physical objects from 3D data models by successive addition of material layer by layer. The input model can be obtained from scanning methods, applying reverse engineering or by CAD models from a 3D modelling software. These technologies have an active participation in RT processes to produce directly tools or patterns to obtain the tools in a secondary process [12, 13]. The integration of AM and RT fields introduces a new trend of tooling practice that has been explored with an active impact on the engineering [1, 9].

Since the 80s, the AM technologies have been used in the product development process. The rapid technological development promoted the use of different terms according to the application field or trademarks. Therefore, there are in the literature and other frequently used terms are rapid prototyping, solid free-form fabrication, additive layer manufacturing, 3D printing and biofabrication for AM biomaterials [12–15]. To facilitate the communication, ISO/ASTM 52900:2015 promotes a standardization terminology for additive manufacturing. RT plays an important role in tooling development, reflecting the evolution of AM not just by reducing lead time but also by improving tool quality and complexity limits [1, 8, 13].

International competition and market globalization require the adoption of new practices of product and process development. One of the most time-consuming and costly phases in process development of a new product is related to tooling for prototypes and components. The application of direct RT processes assists this process, dispensing the production of moulds or patterns, reducing the raw material, energy, time and costs [1, 12].

The application of these techniques has applicability potential particularly in the manufacture of special or customized products, without the use of expensive tools and with shorter manufacturing times and costs. The reproducibility and precision computerized control of the design has numerous advantages for responding to market segments [2, 3, 8]. In addition, the manufacturing method of these technologies enables the production of 3D forms which are impossible to obtain by conventional technologies. In the RT field, this characteristic brings the possibility to produce, for example, new cooling channels systems, reducing cooling times and increasing significantly the number of parts production. One of the limitations of AM is the mechanical properties that are still far from those obtained with conventional processes [14]. Thus, the possibility of making complex near net shaped 3D structures impossible to obtain by conventional technologies, as well as the costs and times of construction are distinctive characteristics of AM [14, 16, 17].

There are many techniques to perform RT by AM using different materials (metal, polymer or ceramic) in different physical states (solid, liquid or paste). The part resolution depends not only on the material properties but also on specific characteristics of the process and machine. The AM processes are classified in seven categories: (i) binder jetting; (ii) directed energy deposition; (iii) material extrusion; (iv) material jetting; (v) powder bed fusion; (vi) sheet lamination and (vii) vat photopolymerization.

1.2.1 Metal Tools

In 1990, the first AM 3D metal part by powder bed fusion process using metal powder formulation of copper, tin and Pb-Sn solder was reported. Different metals and metal composites are used in AM techniques (e.g. stainless steel, steel, bronze, nickel-chromium-based superalloy, copper, iron, cobalt, chromium, nickel, titanium and blends). The most common form is powder but is also used as sheet [18].

Although the most frequent tools are associated with injection moulding, there are other processes that may benefit from RT like casting process, sheet metal forming and forging dies. These processes involve high temperature and mechanical resistance. The metal tooling is more adequate to these conditions [8]. The AM metal processes can contribute to the reduction of tooling time, replacing conventional processes to direct produce products. These techniques can also manufacture more optimized and complex geometries than other conventional technologies. Now, automotive, electronic and medical metal products manufacturing can include low volume production, customized parts and reduced investment costs.

1.2.1.1 Powder Bed Fusion

Powder bed fusion processes apply a thermal energy by an infrared laser beam onto a powder bed material. Selective laser sintering (SLS) is the most used technique compatible with different types of materials (plastic, metal, and ceramics). A thin fusible powdery layer is sprayed into the building platform. The laser beam hits the cross-section powder of one slice of the part allowing a partial fusion of powder interfaces. The building platform descends and a new layer is spread to repeat the process for all part slices. During the print process, the non-sintered powder acts as support material and is also removed. A porous part is obtained with a less accuracy but a higher mechanical resistance than SL parts [19].

In case of the use of metal materials, the technique takes the designation of selective laser melting (SLM) because of the melting of the metal powder. Another process frequently used is direct metal laser sintering (DMLS) or direct laser melting (DLM). Metal powders like iron are coated with a thermoplastic binder which is selective melted. Post-processing steps such as metal infiltration could be also done to increase the densification of parts. SLM is adequate for the production of small moulding inserts with some geometrical complexity and structural resistance [18]. This technique allows the direct production of tooling and components such as injection moulds and copper electrical discharge machining (EDM) electrodes for injection moulding and forge dies [1, 18]. Metal powder moulds infiltrated with copper have also been produced by SLM. Hard tooling solutions were capable to resist a thousand of cycles [20, 21]. These AM processes can produce complex moulds with the possibility to integrate conformal channels which provide a reduction of injection cycle times. Other applications have been reported like jigs for welding and die inserts for casting [22, 23]. The composition of the material, the powder size, and the sintering conditions affect the accuracy and the surface quality. Small powder sizes provide better results but difficult the post processing powder removal process [24].

1.2.1.2 Binder Jetting

After SLS, the three-dimensional printing technique or 3D printing (3DP) was patented. This process, now known as binder jetting, is based on deposition of material (liquid bonding), with ink-jet type printer head, onto a powder bed material. The binder acts as an adhesive, which is sprayed selectively on a print area in each layer of powder, bonding all layers together [1, 18, 25]. Although, currently, the term 3D-printing is related to all additive techniques, originally referred only to a particular

AM technique 3DP. A great number of materials can be processed since synthetic polymers, ceramics, metals, and composites exist. However, it is necessary for an adequate selection of an adhesive agent (binder). The binder selection and process parameters optimization are the key points of this approach. There are studies of different possible powder-binder combinations in literature, but an extensive optimization is needed to develop good quality parts. The design accuracy and resolution can be achieved through the control of powder characteristics and distance between pores.

Metal powder layers are bonded by an aqueous-based or thermoplastic binder to obtain an adequate mechanical resistance for depowdering after printing the parts. The powder around the printed parts acted as support material and allow the assurance of the dimensional stability of printed parts. The critical issue for materials is the mechanical properties of porous structures. The application of resorbable polymer infiltration or heat treatment can increase the strength and toughness. This designated 'green part' can be sintered and subsequently infiltrated with a second metal to merge the metal particles creating a dense part with a better mechanical performance. SLM and 3DP of metal tooling allow the production of functional tooling and a production quantity over 500 units [8].

1.2.1.3 Sheet Lamination

Sheet lamination process consists in the manufacture of parts by successively adding layers of bonded sheets together and cutting according to a projected geometry by a laser beam. Laminated object manufacturing (LOM) is one of the oldest AM techniques that originally worked with paper and a thermoplastic adhesive to bond the sheets. This technique applies successive compression through a preheated roll which activates a heat sensitive adhesive bonding the sheet layers. Each layer is cut by a laser beam to obtain the contour of the slice cross section of a model. All elements that are not part of the printed model are sliced into squares and triangles to facilitate the waste removal of the constructed physical model. The surface of the build platform descends and the process is repeated layer by layer [1]. The resolution of printed parts is determined by the thickness of the raw sheet material used.

This technique was extensively used for prototyping but the introduction of new AM technologies led to a decline in the use of this technology. The technological developments involving the use of other build materials (metal, plastic and composites) stimulated the interest of application in the production of tools. To ensure the structural integrity of printed parts, other cutting strategies beyond the adhesive binder like thermal bonding, clamping and ultrasonic welding were developed. Ultrasonic additive manufacturing (UAM) is a hybrid AM technology with sheet lamination technique that welds sheets or ribbons of metals by an ultrasonic welding. CNC milling is successively used to remove the unbound metal during the addition of material of print process. The process uses metals such as aluminium, copper, stainless steel and titanium [18, 27].

1.2.1.4 Directed Energy Deposition

Directed energy deposition is one of the most recent AM processes based on the application of thermal energy (e.g. laser, electron beam or plasma arc) by melting materials that are deposited into the surface platform, where it solidifies. Different types of materials can be used but the most common is metal (e.g. cobalt chrome and titanium) in the form of powder or wire. It is similar to the material extrusion based process, depositing the fused material layer by layer until the part is completed. For complex parts, it is used to support material or a multiaxis deposition head. Direct energy deposition machines were developed and other designations are used as laser engineered net shaping for metal powder, directed light fabrication, direct metal deposition or 3D laser cladding [18, 26].

1.2.1.5 Material Extrusion

The presented AM processes are the most used technologies for metals. The printed parts have high quality, are functional, and can have a good mechanical performance. However, the methods based on heating source sometimes promote poor surface finish and are not adequate for fine spanning lattice structures. Material extrusion based processes known as robocasting or direct ink writing have been explored to overcome these limitations. Robocasting can deposit a slurry of powdered metal material in the form of filament through a fine nozzle. The material is deposited layer by layer to fabricate 3D parts. These parts are then sintered to create high densities parts. Some metal materials formulation with steel alloys and composites have been explored but more developments are needed in this area [28]. Hybrid structures of electro-ceramic and metal electrode compositions manufactured by this technology have been demonstrated the potential for electronic applications [25].

1.2.2 Polymer Tools

Given the generic characteristics of polymer materials, the tooling using polymers is more suitable for processes with lower loading requirements like thermoforming and fibre forming. However, different polymer tooling have been applied in technologies with more demanding processing conditions (e.g. injection moulding, casting process). As polymers are available in different forms (e.g. powder, filament, liquid), there are many possibilities to produce tooling by AM [8, 12].

1.2.2.1 Vat Photopolymerization

Vat photopolymerization, mostly known as SL was the first additive manufacturing technique in the market in 1987 by 3D System. It is based on a photopolymerization process. A liquid photopolymer in a vat is selectively cured layer by layer by light-activated polymerization. This technique is limited by the use of photopolymerizable materials and involves the realization of post-processing (postcure with UV light) for a better consolidation and for increasing the mechanical resistance. The most common materials are acrylic or epoxy resins. For complex geometries, support structures with the same build material are created. A small point contact of these structures provides an easy removal. For biomedical field, only synthetic (e.g. polycaprolactone (PCL)) and natural (e.g. alginate and chitosan) biopolymers can be processed with this process [12, 29, 30].

SL techniques have been used to produce inserts and mould core and cavity for injection moulding. An assembly of aluminium blocks with SL cavity and core inserts was used to inject thermoplastic parts. Tooling made by SL proves to be a RT alternative for use in short-run injection moulding around hundreds of parts [31]. Other studies conclude that the great advantage of the use of SL in RT is the time and costs of tool building. High-dimensional accuracy is reached but poor resistance and thermal fatigue sometimes result in structural integrity problems [18, 32]. SL of epoxy resins has been applied to the production of extrusion dies and injection moulds. These tools can inject small production runs from different polymers (e.g. polyurethane, low-density polyethylene and PCL). Investment casting process can benefit from SL castable resins to produce detailed metal parts for jewellery, metalworking and engineering applications [1].

1.2.2.2 Material Extrusion

Material extrusion, also called as Fused Filament Fabrication (FFF), consists on the production of 3D parts, layer by layer, by dispensing material through a nozzle or orifice. By the addition of successive layers of material, the melted filament is deposited creating a complete printed part. During the solidification, each layer connects to the previous layer. Support material is printed to assure the stability of the layers and removed by mechanical tools or in a chemical liquid in a secondary operation. Thermoplastic materials are broadly used, with the emphasis on acrylonitrile butadiene styrene (ABS), polylactic acid (PLA), PCL and polyamide (PA) [16, 33]. In the last years, nozzle-based (material extrusion) AM systems, particularly thermal extrusion, have shown a high continuous growth in comparison to the laser-based systems. This results in the appearance of various desktop techniques at a reduced cost. Accessibility and operation simplicity led to an increase in the investigation with these fabrication strategies. The spatial resolution is limited by the diameter of the extrusion nozzle and, consequently, by material rheology characteristics. These polymer 3D parts are still used as conceptual prototypes. The lack of strength and functionality restricts the type of applications. The introduction of engineering materials (e.g. polyoxymethylene (POM) and polyether ether ketone (PEEK)) and composites confers to the parts mechanical properties with potential for industrial application. The use of these materials promotes the possibility of obtaining parts with complex geometry impossible to obtain with conventional injection moulding techniques. A lot of different filaments have been introduced, including polymer matrix composites with particles, fibre or nanomaterial reinforcements. Metal materials like copper, bronze, steel, ceramic materials like glass, thermoplastic elastomer and carbon or graphene nanomaterials are some examples of reinforcement materials [12]. FFF of PLA proves to be adequate to produce casting cores for metal casting with higher accuracy and better surface quality [30, 34].

Robocasting can be considered as another material extrusion process where a filament of suspension or paste is extruded from a nozzle layer by layer. Thermoplastics (e.g. PLA and PCL) and elastomers (silicone rubber) are some polymeric materials used. A nozzle of lower diameter allows the production of parts with a higher resolution with complex structures, useful for applications as specific filters or membranes [35].

In terms of applications, these technologies have already been successfully applied in biomedical applications for tissue engineering like scaffolds for bone defects or other tissues and organs such as ears, aortic valves or cartilage constructs. The electronic field is also being explored, printing electronic devices like piezoresistive sensors. This AM technology also offers the possibility to manufacture aerospace complex components. A 3D extrusion method was used to produce an inlet guide vane of polymer reinforced with carbon fibre. Engineering materials such as PEEK were also applied in aerospace applications with high performance [12].

1.2.2.3 Material Jetting

More recently, new AM techniques based on material jetting process (e.g. PolyJet) were developed. The process consists in the selective deposition of droplets of build material (photopolymer or waxes) [12]. The polymerization occurs in each layer by ultraviolet (UV) light projection. The material is jetted in a build platform and solidifies (cure or harden) layer by layer by UV light. A suitable viscosity and the ability to form drops are required for the build materials. A gel-like support material is also deposited in the same way to build materials and is removed in post-processing (e.g. water jet or by hand). The process allows the control of the material composition of the product by jetting different materials combining different 3D printing materials in the same 3D printed part and in the same print job [36]. Some of these parts can have characteristics similar to parts obtained by a conventional bi-injection moulding process.

Functional parts with high resolution can be produced with this technique. The engineers can produce prototypes and use as a tool for a better communication and design validation tool. Prototypes are realistic, they can have moving parts and are cheaper when compared with CNC machining. The technology is being applied in low volume productions without tooling. It can still be used to quickly create moulds or inserts for the production of small parts. RTV silicone was injected into a mould obtained by PolyJet to create a keypad in a short time [36]. This technology defines a poorer surface finish compared to SL but its flexibility in terms of material mechanical properties and building speed brings a place in the development of production tooling.

FFF and PolyJet techniques have been applied for the production of patterns for investment casting and thermoforming moulds with shorter delivery time than the conventional methods [37, 38].

1.2.2.4 Binder Jetting

The binder jetting processes previously described are also compatible with polymeric materials (e.g. Polymethylmethacrylate (PMMA)). One of its advantages is the flexibility in terms of materials and the room temperature processing environment. However, the mechanical resistance of printed parts is weak compared with other AM technologies. Although this limitation, the 3DP is applied in the production of RT [12]. Investment casting models made by 3DP of PMMA are possible to obtain without tools and with complex geometry. The delivery time of these tools is reduced and the wax infiltration, similar to the conventional wax models, can be made. The typical roughness of the parts produced by this technology can be smoother with wax infiltration. Epoxy resins can also be applied to increase the strength of the printed parts.

1.2.2.5 Powder Bed Fusion

In the case of powder bed fusion processes, the most applicable materials are PA, polystyrene, polypropylene, and composites. SLS does not need the production of support structures because the powder around the part supports the surfaces. This factor allows creating tooling with more complex geometries. Perforated mould for fibre forming process was made by SLS [8]. SLS of PCL models is used as an alternative of wax patterns for investment casting. A greater dimensional accuracy and improved strength can be obtained [9].

1.2.3 Ceramic Tools

Ceramic materials are mostly used in AM and RT technologies in the form of powder [8]. The critical issue for materials is the mechanical properties of porous structures. The AM ceramic materials involve normally thermal treatment or infiltration treatments to ensure the adequate mechanical performance. The resorbable polymer infiltration and the use of resorbable glassy materials can increase the strength and toughness of the ceramic material but needs an extensive study of process-property optimization [30].

1.2.3.1 Binder Jetting

Plaster and sand are the most common commercial material available for Binder jetting processes. The AM of ceramic materials has been applied as RT for ceramic processing processes (e.g. slip casting) and metal casting processes. 3DP plaster moulds can be directly produced by these techniques. Silica sand and furan/phenol resins are processed by 3DP to manufacture sand moulds and cores for lost-mould casting methods. Complex and large tools can be fabricated directly without moulds. 3DP tools are suitable for conventional castable materials like steel, aluminium, iron, brass, magnesium or metal alloys [11].

A sand core for a turbine wheel was produced by 3DP in a single piece. Compared with conventional methods, 3DP allows not only the reduction of tooling costs but also eliminates the core assembling step. 3DP tooling for the vacuum-assisted resin transfer moulding (VARTM) shows less accuracy than CNC machined tooling but is more advantageous in terms of costs and time. 3DP was used to print plaster moulds for carbon fibre moulding. These moulds present thermal stability and are water soluble. Polyester or epoxy resins are used to reinforce the moulds and improve the surface finish caused by intrinsic porosity [39]. 3DP is also used in moulds for casting any castable alloy. This technique is suitable for the production of complex moulds with a gating system and internal cores for functional metal parts [1, 9].

1.2.3.2 Powder Bed Fusion

Powder bed fusion processes are also applied in ceramic tooling. Silicon carbide (SiC), aluminium oxide (Al_2O_3) and zirconium silicate ($ZiSiO_4$) ceramic parts have been investigated for the production of tooling by a direct method. SLS was applied to produce investment casting shell for impeller wheel [40].

SLS of aluminium was used to produce functional and structural ceramic and ceramic cores. Ceramic cores contribute to intricate cooling passages of investment cast turbine components. The application of AM technologies eliminates long lead time of traditional tooling. A direct method of SLS of ZiSiO4 powder was investigated to produce a ceramic shell to be applied in investment casting [41].

1.2.3.3 Material Extrusion

The material extrusion process is also used to AM ceramic parts. Robocasting is capable to produce ceramic parts for RT. Slurries with highly concentrated colloidal suspensions are used in this technique. Alumina components can be fabricated by robocasting [30].

1.3 Indirect AM Rapid Tooling Processes

Indirect AM RT process focuses on the production of a master model for small volume productions. Typically, soft tooling is produced from silicone rubber tools. Master models created by AM technologies are used to fabricate this soft tooling. Casting or moulding processes of plastics, ceramics or metals use this type of tool. Plaster or ceramic moulds can be also created for micro investment casting of metals [30].

Soft tooling is commonly obtained by silicone moulds, castable resin and RTV process. The soft tooling involves low costs and it is applied to do low volume production or even just one part (e.g. prototype). However, it is possible to produce hard tooling using durable materials and indirect methods such as metal spraying, cast metal tooling, and keltool tooling. Normally, an intermediate step is made to produce indirect tooling. A master is created to be copied many times to fabricate strong tools able to produce a higher volume of parts [1, 9].

The indirect methods start with the fabrication of patterns. The AM technologies contribute especially to rapid manufacture of these master models. Consequently, the mould quality depends on patterns quality. Studies report the use of AM technologies to produce patterns for investment casting. The patterns obtained by SLS and 3DP need to be improved for obtaining better surface quality. The decrease in porosity can be achieved by using small powders sizes or by using infiltrants [19].

1.3.1 Metal Tools

Wire, powder or liquid (by melting) forms are the most used state form in indirect methods that involve metals. Metal tools can be manufactured by spraying low melting alloys such as lead or tin alloys. The processes may begin with a material in a yarn form that is melted and sprayed with a compressed gas. Metals can be also manufactured by casting processes to produce aluminium or zinc tools for the production of few pieces or steel tools for a high volume of pieces. In some cases, there are applications of some metal inserts, made by steel, copper or aluminium alloys, or surface coatings with copper or nickel to improve the mechanical and thermal properties (e.g. thermal conductivity) [1, 12, 30].

1.3.1.1 Metal Spraying and Deposition

Hard tooling may be obtained by an indirect RT way such as metal spraying process. Studies report the application of spray metal moulds applied in low-pressure processes (e.g. rotational moulding and vacuum forming). Depending on materials, the metal spray tooling can achieve adequate abrasion properties for high volume production. LOM, SLS, FFF and SL techniques are able to produce masters from different materials such as wood, metal, PCL, ABS and resin. The production of metal spraying moulds consists in the application of a coating in patterns by spray metal (e.g. nickel, tin-zinc or steel) with compressed air electric gun. After it is backed with metal filled epoxy resin to prevent the formation of cracks. A hundred injection moulding parts can be obtained by these tools. Low volume production such as prototypes can be also supported by spray metal tools. Aerospace industry uses electrodeposition process to obtain a nickel shell [1, 9]. Intermediate tools from SL patterns were used to create a metal master tool by metal spraying process for fabricated spin casting moulds [42].

1.3.1.2   Metal Casting

Some injection moulding and die casting tooling are produced by investment casting. Earlier some AM technologies useful for the production of patterns were mentioned. The tool life depends on the material composition used. Cast metal tooling of aluminium and zinc are used for the production of prototypes and steel is used for high volume production [9]. FFF was also used to produce complex wax patterns for investment cast aluminium mould inserts with short delivery time.

1.3.1.3   3D Keltool

3D Keltool process (3D Systems Inc.) has been used to produce hard tool inserts. SL master pattern is used to produce silicone mould by RTV. Metal powder coated with a plastic binder is poured into the silicone mould and green parts are obtained. Thermal treatments and cooper infiltration are applied to create dense parts with high hardness and durability [9].

1.3.2   Polymer Tools

Since many indirect processes involve castable materials, the liquid form of the materials is the most requested. In the case of plastics, polyurethane and resins are the most versatile materials in terms of properties and therefore the most used. With these materials, it is possible to obtain mechanical and thermal properties similar to different thermoplastics (e.g. ABS, PA). It is possible to manufacture a few tools with low cost and characteristics like tools obtained by conventional processes such as injection moulding. The resins can also be used with reinforcements, like aluminium powder, to increase the mechanical performance (e.g. thermal conductivity) [1, 12, 30].

1.3.2.1 Silicone Moulding

Silicone moulding is one of the most used processes to produce polymer soft tooling. It is used to manufacture wax patterns, plastic parts or metal parts with a low melting point. The pattern is produced by AM technologies and flexible silicone mould is created from the pattern. After cured, the silicone mould is cut and the pattern removed. A variety of polyurethane parts are produced by vacuum casting for low volume production (up to 20 units) [9]. This method has the advantage of creating inexpensive moulds for small and large parts. Studies report the successful application of silicone rubber tools for the production of wax patterns. The advantages of silicone rubber in terms of flexibility and thermal resistance (200°C) lead to its wide use such as tooling in many conventional processes. RTV process with and SL models are frequently used to make silicone rapid tools [42]. These types of tools are also used for reaction injection moulding (RIM), wax injection tooling and spin casting [1].

1.3.2.2 Resin tooling

Epoxy moulds have a larger lifetime compared with silicone moulds. Epoxy resin tooling is made from SL, SLS or LOM models. The models are coated with a release agent (e.g. clay or plaster) and it is produced by pouring the liquid resin over the models in the mould box. Two parts of a mould are made separately. The use of epoxy moulds for low temperature and pressure processes such as RIM is especially suitable for the production of few thousands of parts [9].

Metal reinforcement is used in epoxy resin to improve the mechanical and thermal performance (thermal conductivity) of moulds. During the production of the resin, moulds can incorporate cooling channels adding copper coils. Plastic and wax injection moulding uses this type of tools. Cast resin tools have the ability to produce 100–300 parts [1, 42].

1.3.3 Ceramic Tools

Non-technical and technical ceramic powders and slurries are also used to produce tooling, especially patterns for casting process and technical components. Plaster and different materials formulations like silica with a binder (urea formaldehyde resin), zircon sand and alumina-silica, zirconium silicate, aluminium oxides, and silica are some examples of tooling compositions [1, 9, 12, 30].

1.3.3.1 Castable Patterns

3DP is used to produce patterns to make the plaster models for ceramic processes. These tools are useful for the manufacturing process of traditional ceramic materials.

Conventional slip casting processes use plaster moulds made by 3DP patterns. In addition, powder compaction in presses and isostatic forming can also use this RT for structural ceramics production of technical products (chip carriers, spark plugs and cutting tools). The possibilities to do patterns reproducible without specialized labour are some of the benefits for the ceramic industry. The use of AM techniques contributes not only to short lead times but also for the technological development of this industrial sector [1].

Investment casting ceramic tools are fabricated using 3DP processes to do patterns. Investment casting ceramic shells were fabricated covering the patterns with a ceramic shell. For a better consolidation, these shells were sintered. This ceramic tooling promotes a good surface quality parts and fine features [1]. Another ceramic tool technology designed by Shaw technology is used for the precision casting process. A ceramic material in the form of slurry is poured into a box which has the pattern. Then the slurry hardens, the pattern is removed and the tool is sintered. 3DP and SL technologies were used to fabricate patterns. In case of 3DP, a wax pattern that was destroyed in the sintering process was produced [42].

1.4 Other RT Technologies

Although the RT concept is majorly associated with the additive manufacturing processes, some techniques have been proposed using different technologies. A common framework can be established for all CAD-based manufacturing principles, with the tool parts developed through automated methods in reduced time. Often, the developed tools require manual assembly or finishing operations.

In such a way, conventional tooling development processes such as CNC cutting and machining processes and EDM feature a possible use as RT processes. In many cases, these processes are used to manufacture prototype tools to be used as bridge tooling. Common base assemblies can be used to accommodate mould inserts to define moulding core and cavity. The RT with the use of CNC machining is used in the development of these inserts. In some simpler manufacturing processes, the rapidly developed parts can be used alone for moulding operation.

1.4.1 Rapid Machining Tools

The use of machining technologies in RT applications finds its most important use in the development of prototype moulds or bridge tools. Due to its importance on the product development industry, most of the examples feature the development of plastic injection tools. However, this principle can be used for the development of tools for a large variety of manufacturing processes.

Although the machining RT processes use the same equipment used for conventional tool making, the mould assembly strategy and materials used are different, leading to a faster development time at the cost of a shorter tool life.

1.4.1.1 Milling Materials for RT

Milling operations are achievable in a large variety of materials. However, the mechanical properties of the materials have a large influence on the machining parameters, thus, largely influencing the process time.

Due to its great mechanical characteristics, metals are used in machining RT operations. However, most applications condition the material selection not only based on the tool performance but also on the tool manufacturing ease. In such a way, aluminium alloys are mainly used for the development of rapid machined metal tools. The machinability of aluminium alloys is considerably higher than steel, leading to higher cutting speed and less cutting tool wear. Besides, the low density of aluminium, when compared with tool steel, also benefits the parts handling during fabrication. From the performance point of view, the higher thermal conductivity benefits the moulding cooling and heating, leading to a potential simpler cooling system. The most common aluminium alloys used in machined RT are AlZnMgCu: 7000 series in higher performance demanding tools and AlMgSi: 6000 series in lower cost tools.

Mild steel and different types of mould steel types such as pre-hardened steels and corrosion resistant steels are also used when the parts require a gloss or clear surface finishing or higher wear resistance. However, the use of these materials leads to longer manufacturing time and often requires coating protection.

Apart from the operation using metal, there are a number of board materials available for RT CNC machining. Some of the most relevant commercial available examples are Ureol, Necuron, Araldite and Cibatool. These polymeric resins, mainly epoxy or polyurethane based, have been developed both for prototype development and for RT applications. The use of these materials allows a simple one step of the tool while maintaining a high strength and excellent thermal resistance. Although the use of these materials offers larger geometry limitations, as EDM is not possible, they allow developing tools with no internal cooling required and little or no finishing required [20].

1.4.1.2 Insert Moulds

The use of machining operations in RT finds its main operation in the development of simplified mould structures to achieve a lower cost and shorter lead time. The tools use a common master unit die (MUD) with standard tooling components where a cavity and core inserts are placed.

The tools MUD is typically fabricated using steel alloys and standard tool components. These MUDs are developed and fabricated to be available in different sizes and allowing different moulding features such as guidance systems, temperature control systems, insert retainers and possibility to include slides. As these bases can be pre-developed, they do not influence the time needed to complete a dedicated mould fabrication as it is only influenced by the cavity and core inserts manufacture.

The machined RT using insert moulds is mainly performed for the development of prototype moulds, both for low volume manufacture or as bridge tooling. Several tool manufacturing companies use this technique to produce low-cost rapid development tools using aluminium alloys.

The use of resins allows to speed up the machining time as larger step downs or even one-step direct milling is possible. On the other hand, these materials are not compatible with EDM, limiting the achievable geometry. Geometries with undercuts, deep slots or sharp inner edges are not feasible for machining operations. To overtake these difficulties, the cavity and core inserts can be divided into smaller parts, using a base part and small inserts. The assembly of simpler parts allows achieving higher complex geometries, although they may lead to material burrs on the inserts interface. This strategy can be used both with fixed and movable inserts to help part demoulding [43].

The milling of some small inserts may be performed using sacrificial supports to allow the tool approach under different orientations. This strategy, in addition to the use of multiple cutting tools, leads to better and faster results than most rapid manufacturing processes [44].

1.4.1.3   Space Puzzle Moulding

The use of large inserts and MUDs in a mould assembly enables to achieve fast developed tools at a reduced cost. However, the geometric complexity of these tools is limited to machinability issues. Undercuts are possible using slides, at the cost of longer development time and additional cost.

An alternative approach to reach higher complexity parts is the use of multi-piece tools such as space puzzle modelling. The mould is divided into multiple parts, each with a different parting direction. The mould pieces are hand-loaded into a base and mounted on a moulding machine. The moulding and material cooling are performed while the mould pieces are accurately and securely clamped into the holding device. Finally, each mould piece is hand-removed from the mould base to release the moulded part.

From both the tool development and the moulding process points of view, the tool should be designed to have the minimum number of mould pieces that can form the cavity of the part, while ensuring a possible disassembly in the part ejection process [10, 45].

1.4.1.4   Patternless Casting

A different field of application of the machining RT technologies is the development of tools for metal casting operations.

Conventional casting processes take advantage of patterns for the mould development. While this approach is economically interesting for the fabrication of medium to large volumes, it is not well suited for the fabrication of unique parts or small batches. As an alternative, milling operations can be used as RT technology for the development of casting moulds. This approach not only lowers the production cost for low volume production but also speeds up the time to first part.

The patternless process implies the use of CNC milling operation for the definition of the part cavity in a hardened block of moulding sand. This method uses second-generation moulding sands, containing synthetic resins like furan moulding sands or phenol formaldehyde moulding sands. During the operation, the milled moulding sand is immediately drained to prevent damaging the mould surface. After milling, the surface of the mould cavity is coated to prevent the metal penetration [46].

1.4.2   Laminae Tools

An alternative solution to the assembly of milled blocks has been proposed using plate configuration materials. These alternatives have been introduced not only to allow a faster or simpler mould development but also to allow the inclusion of free shape cooling channels inside the mould.

1.4.2.1   Lamination of Metal Sheets

The first tooling development process involving the use of metal sheets involves the lamination of plates. The process involves the cutting of cross section in metal sheets, assembly, clamping and bounding of the parts and milling and finishing operations to conclude the mould.

Each metal lamina is laser cut and manually deburred. The parts are assembled and pressed between two metal plates. The tool development uses coated aluminium sheets to allow bounding at a relatively low temperature. The flux coating works as a bounding material as it melts without significant tool material shrinkage to affect the thinner features. Afterward, core and cavity moulding surfaces are finished using standard techniques. Thus, precision and surface quality are comparable to conventional tool making. Major difficulties are found in small slopes as they result on thin layer edged and large surface area from a single lamina. Surface quality and hardness can be improved by coating [47].

1.4.2.2 Profiled Edge Laminae Tools

The possibility of designing a tool based on thick material layers has been developed as an alternative RT technology. A process called profiled edge laminae (PEL) tooling involves assembling of laminae sheets, each one with a uniquely profiled and bevelled top edge. The layers are put together vertically in a precise and repeatable manner with a common bottom edge and side edge supported in two precisely machined reference planes. The layers are held together under pressure in a simple bolt clamping system and/or bonded by brazing into a rigid tool [10, 48].

PEL tooling offers advantages over CNC machining of a solid billet of material, including the ability to route conformal heating/cooling channels and process sensors or incorporate special features such as vacuum lines. The mould tool accuracy is both influenced by the cut of each lamina and by the slicing algorithm used during tool development [49].

The tool definition is defined by the assembly of the layers. Each layer top edge is cut from a CAD model in a CNC controlled cutting process. Abrasive water jet cut, laser cut, plasma cutting or wire EDM can be used for the cut of the top edges. The selection of each layer thickness is performed according to available sheets to avoid unnecessary milling operations. The tool assembly may use various layer thickness to better adapt to the tool surface. The use of five-axis cutting systems allows the cut of bevel edges and better defines the tool surface. The definition of PEL tools may use different materials in plate form. However, metal plates from aluminium and steel alloys are mainly used with thickness from 1 to 25 mm [49].

References

1. A. Equbal, A. Kumar Sood, M. Shamim, Rapid tooling: a major shift in tooling practice. J. Manuf. Ind. Eng. **14**(3–4), 1–9 (2015)
2. P. Hilton, P. Jacobs, *Rapid Tooling: Technologies and Industrial Applications* (Marcel Dekker, New York, 2010)
3. P. Vasconcelos, F. Lino, R. Neto, M. Vasconcelos, Design and rapid prototyping evolution, in *RPD 2002—Advanced Solutions and Development Conference* (2002)
4. T. Andrew, Development of an expert system as applied to rapid tooling techniques for injection molding. Master's thesis, Lehigh University (2005)
5. E. Tackett, *Rapid Tooling* (Saddleback College Advanced Technology Center, 2012)
6. K. Karunakaran, S. Suryakumar, A. Bernard, Hybrid Rapid Manufacturing of Metallic Objects. 14èmes Assises Européennes du Prototypage & Fabrication Rapide (2009)
7. N. Hopkinson, R. Hague, P. Dickens, *Rapid Manufacturing: An Industrial Revolution for the Digital Age* (Wiley, West Sussex, 2007)
8. G.N. Levy, R. Schindel, J.P. Kruth, Rapid Manufacturing (LM) technologies, state of the art and future perspectives. CIRP Ann. Manuf. Technol. **52**(2), 589–609 (2003)
9. A. Rosochowski, A. Matuszak, Rapid tooling: the state of the art. J. Mater. Process. Technol. **106**(1–3), 191–198 (2000)
10. Wohlers, Wohlers Report 2006–2016, 3D Printing and Additive Manufacturing State of the Industry, Wohlers Associates, Denver (2006–2016)
11. P. Yarlagadda, L. Wee, Design, development and evaluation of 3D mold inserts using a rapid prototyping technique and powder-sintering process. Int. J. Prod. Res. **44**(5), 919–938 (2006)

12. X. Wang, M. Jiang, Z. Zhou, J. Gou, D. Hui, 3D printing of polymer matrix composites: A review and prospective. Compos. Part B Eng. **110**, 442–458 (2017)
13. ISO/ASTM International, ISO/ASTM 52900:2015—Additive manufacturing—General principles—Terminology. ISO/ASTM International (2015)
14. G. Gmeiner, U. Deisinger, J. Schnherr, J. Stampfl, Additive manufacturing of bioactive glasses and silicate bioceramics. J. Ceram. Sci. Technol. **6**(2), 75–86 (2015)
15. L. Pires, Biocermicos e Biovidros para prototipagem 3D: propriedades e formulaes. Masters thesis, University of Aveiro (2011)
16. Q. Yao, B. Wei, Y. Guo, C. Jin, X. Du, C. Yan, J. Yan, W. Hu, Y. Xu, Z. Zhou, Y. Wang, L. Wang, Design, construction and mechanical testing of digital 3D anatomical data-based PCLHA bone tissue engineering scaffold. J. Mater. Sci. Mater. Med. **26**(1) (2015)
17. J. Ferreira, A. Mateus, Studies of rapid soft tooling with conformal cooling channels for plastic injection moulding. J. Mater. Sci. Mater. Med. **142**(2), 508–516 (2003)
18. W. Sames, F. List, S. Pannala, R. Dehoff, S. Babu, The metallurgy and processing science of metal additive manufacturing. Int. Mater. Rev. **61**(5), 315–360 (2016)
19. A. Pouzada, Hybrid moulds: a case of integration of alternative materials and rapid prototyping for tooling. Virtual Phys. Prototypes **4**(4), 195–202 (2009)
20. D. King, T. Tansey, Alternative materials for rapid tooling. J. Mater. Process. Technol. **121**(2–3), 313–317 (2002)
21. A. Armillotta, R. Baraggi, S. Fasoli, SLM tooling for die casting with conformal cooling channels. Int. J. Adv. Manuf. Technol. **71**(1–4), 573–583 (2014)
22. S. Campanelli, N. Contuzzi, A. Angelastro, A. Ludovico, Capabilities and performances of the selective laser melting process, in *New Trends in Technologies: Devices, Computer, Communication and Industrial Systems* (2010), pp. 233–252
23. E. Pessard, P. Mognol, J. Hascot, C. Gerometta, Complex cast parts with rapid tooling: rapid manufacturing point of view. Int. J. Addict. Manuf. Technol. **39**(9–10), 898–904 (2008)
24. J. Milovanovic, M. Stojkovic, M. Trajanovic, *Metal Laser Sintering for Rapid Tooling in Application to Tyre Tread Pattern Mould. Sintering - Methods and Products* (InTech, 2012)
25. J. Smay, S. Nadkarni, J. Xu, Direct writing of dielectric ceramics and base metal electrodes. Int. J. Appl. Ceram. Technol. **4**(1), 47–52 (2007)
26. I. Gibson, D. Rosen, B. Stucker, *Sheet Lamination Processes* (Additive Manufacturing Technologies, Boston, MA, 2010), pp. 223–252
27. R. Friel, R. Harris, Ultrasonic additive manufacturing—a hybrid production process for novel functional products. Procedia CIRP **6**, 35–40 (2013)
28. M. Yetna NJock, E. Camposilvan, L. Gremillard, E. Maire, D. Fabrgue, D. Chicot, K. Tabalaiev, J. Adrien, Characterization of 100Cr6 lattice structures produced by robocasting. Mater. Des. **121**, 345–354 (2017)
29. D. Kuznetsova, P. Timashev, V. Bagratashvili, E. Zagaynova, Scaffold and cell system-based bone grafts in tissue engineering (review). Sovrem. Tehnol. Med. **6**(4), 201–211 (2014)
30. T. Hanemann, W. Bauer, R. Knitter, P. Woias, Rapid prototyping and rapid tooling techniques for the manufacturing of silicon, polymer, metal and ceramic microdevices, in *MEMS/NEMS*, Boston, MA (2006), pp. 801–869
31. A. Do, P. Wright, C. Sequin, Latest-generation SLA resins enable direct tooling for injection molding. Soc. Manuf. Eng. **5**(3), 1–15 (2000)
32. V. Beal, C. Ahrens, P. Wendhausen, The use of stereolithography rapid tools in the manufacturing of metal powder injection molding parts. J. Braz. Soc. Mech. Sci. Eng. **26**(1), 40–46 (2004)
33. J. Temple, D. Hutton, B. Hung, P. Huri, C. Cook, R. Kondragunta, X. Jia, W. Grayson, Engineering anatomically shaped vascularized bone grafts with hASCs and 3D-printed PCL scaffolds. J. Biomed. Mater. Res. Part A **102**(12), 4317–4325 (2014)
34. C. Mendonsa, V. Shenoy, Additive manufacturing technique in pattern making for metal casting using fused filament fabrication printer. J. Basic Appl. Eng. Res. **1**(1), 10–13 (2014)
35. E. Malone, H. Lipson, Fab@Home: the personal desktop fabricator kit. Rapid Prototyp. J. **13**(4), 245–255 (2007)

36. M. Javaid, L. Kumar, V. Kumar, H. Abid, Product design and development using polyjet rapid prototyping technology. Control. Theory Inform. **5**(3), 12–19 (2015)
37. O. Marwah, S. Sharif, M. Ibrahim, E. Mohamad, M. Idris, Direct rapid prototyping evaluation on multijet and fused deposition modeling patterns for investment casting. Proc. Inst. Mech. Eng. Part L J. Mater. Des. Appl. **230**(5), 949–958 (2016)
38. C. Hartman, V. Rosa, Benefits of 3D printing vacuum form molds. FATHOM (2014)
39. D. Dippenaar, K. Schreve, 3D printed tooling for vacuum-assisted resin transfer moulding. Int. J. Adv. Manuf. Technol. **64**(5–8), 755–767 (2013)
40. J. Kruth, P. Mercelis, J. Van Vaerenbergh, L. Froyen, M. Rombouts, Binding mechanisms in selective laser sintering and selective laser melting. Rapid Prototyp. J. **11**(1), 26–36 (2005)
41. C. Cheah, C. Chua, C. Lee, C. Feng, K. Totong, Rapid prototyping and tooling techniques: a review of applications for rapid investment casting. Int. J. Adv. Manuf. Technol. **25**(3–4), 308–320 (2005)
42. J. Wang, X. Wei, P. Christodoulou, H. Hermanto, Rapid tooling for zinc spin casting using arc metal spray technology. J. Mater. Process. Technol. **146**(3), 283–288 (2004)
43. N. Volpato, J. Amorim, Systematic to overcome CNC machining limitation in rapid tooling, in *19th International Congress of Mechanical Engineering* (2007)
44. M. Zahid, K. Case, D. Watts, Cutting Tools in Finishing Operations for CNC Rapid Manufacturing Processes: Experimental Studies. Int. J. Mech. Aerosp. Ind. Mechatron. Eng. **8**(6), 1108–1112 (2014)
45. S. Stoyan, Y. Chen, Multi-piece mold design based on linear mixed-integer program toward guaranteed optimality (2010)
46. R. Pastirik, D. Urgela, Device for production of prototype moulds by milling. Achieves Foundry Eng. **11**(si.1), 45–50 (2011)
47. T. Himmer, T. Nakagawa, M. Anzai, Lamination of metal sheets. Comput. Ind. **39**(1), 27–33 (1999)
48. S. Yoo, D. Walczyk, Advanced Design and Development of Profiled Edge Laminae Tools. J. Manuf. Process. **7**(2), 162–173 (2005)
49. S. Yoo, D. Walczyk, An adaptive slicing algorithm for profiled edge laminae tooling. Int. J. Precis. Eng. Manuf. **8**(3), 64–71 (2007)

Chapter 2
Incremental Sheet Forming

2.1 The Role of Incremental Sheet Forming

Incremental sheet forming (ISF) is a sheet metal forming technique where a sheet is formed into the final workpiece by a series of small continuous incremental deformations. The process is controlled entirely by CNC technology and no die is needed as is in traditional sheet metal forming. The removal or simplification of the die in the manufacturing process decreases the cost per piece and improves turnaround time for single parts or low batch production runs. On the other hand, the differences in the forming principle lead to a loss of accuracy and differ the possible achievable part design. Incremental forming processes can be considered as rapid prototyping as they are well suited for small-batch production and rapid production of service parts and may reduce time to market [1, 2].

The main advantages of ISF include a short time to market because parts can be formed through CAD/CAM methods, by directly generating CNC tool path strategies from 3D CAD models. ISF eliminates partially or totally the need for tooling development and manufacture. The process is thus cost-effective for small sized series or even unique parts manufacture. The process allows a high degree of flexibility, being adapted to very different free-form geometries, materials and forming conditions. Changes in part design can be easily and quickly accommodated and the shape complexity has no major throwback on the manufacturing cost. ISF allows producing parts of any size, with the only limitation related to the available hardware. In addition, the formability of materials under the localized deformation imposed by incremental forming is higher than in conventional forming processes [3–5].

Incremental sheet forming processes have been majorly studied since the beginning of the 2000s. With the consistent results that it is now possible to obtain, the process dissemination around the industrial companies is starting. Besides the applications in the prototyping field, ISF can be used for the manufacture of unique parts and small volume production.

© The Author(s), under exclusive licence to Springer Nature Switzerland AG 2019   23
D. Afonso et al., *Incremental Forming as a Rapid*
Tooling Process, Manufacturing and Surface Engineering,
https://doi.org/10.1007/978-3-030-15360-1_2

The ISF process consists of gradually forming a sheet to the desired shape, by following a tool path with a simple tool, defined by CAD/CAM techniques. The process is capable of defining free-form surfaces without the need for dedicated tools. Due to their operation principle and capabilities, these processes fill the role of smart manufacturing processes in sheet metal forming, highly suited for rapid product development and rapid manufacturing.

Major issues with incremental forming processes are related to low accuracy and excessive thinning. The process is also influenced by the rigidity of used machine and by a large amount of springback. However, ISF offers high reproducibility and enables to produce sheet metal parts with a good surface quality and mechanical properties [4, 5].

2.2 Operation Principle

Incremental sheet forming processes consist of gradually forming a sheet to the desired shape, by following a tool path with a simple punch tool. A large variety of ISF principles have been developed in the past years. The process has first been introduced in the 60s, even though it has been on hold due to difficulties in the motion control [6]. Presently, the most settled operation is single-point incremental forming (SPIF). A sheet is peripherally supported by a dedicated backing plate with the part out frame, and a simple punch follows a continuous tool path, forming the blank to the final shape. Other techniques like dual-point incremental forming (DPIF) replace the backing plate by a second motion controlled tool, providing either local or peripheral support and eliminating the need for a dedicated tool. By the opposite, approaches like two-point incremental forming (TPIF) use low cost complete or partial dedicated dies to back the sheet forming process. Figure 2.1 represents the working principle of the SPIF, DPIF with local support (DPIF-L) and with peripheral support (DPIF-P) and TPIF. The first three columns of Fig. 2.1 represent the beginning of the forming processes. The forming tool approaches the blank and starts the forming operation. This process continues while the forming depth increases. The last column of Fig. 2.1 indicatively represents the final part shape with different depth levels of the tool path. Besides this, several other variants exist, including the use of partial or complete low cost dies, use of multiple tools, use of elastic die and the use of special techniques or auxiliary tools to improve formability [3].

All ISF forming processes use a clamped sheet. Both flat and pre-formed blanks can be used. A simple tool, generally spherical, is CNC controlled to follow a tool path against the sheet. The first approach consists of moving the forming tool to a position over the sheet. The following movement leads the tool to the entrance point in a plunge or ramp path. The tool path then continues to deform the sheet, deepening the tool tip along the process according to a predefined pitch. When the final shape

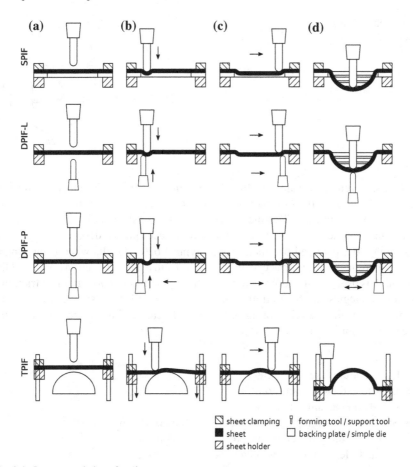

Fig. 2.1 Incremental sheet forming processes

is achieved the tool retracts and moves away from the sheet. The most common forming strategies use countering tool path with either localized plunge between vertical increments or continuous helical movement, based on finishing strategies from milling operations. Both constant z steps and constant scallop height can be used, as represented in Fig. 2.2 by Δz and Δh. While constant z steps are often faster, constant scallop height leads to better part quality, mainly when dealing with low slope geometries. Vertical steps used typically vary from 0.1 to 2 mm, according to material thickness and strength and the accuracy and surface finishing pursuit. Feed rate speed varies from very low speeds as 100 mm/min to high-speed ISF over 12000 mm/min. The forming force, mainly vertical, depends on material thickness and mechanical properties, as well as on the forming parameters. The force may vary from 200 N when forming thin soft materials with small increments to 20 kN when forming thicker and harder materials. Total forming time depend strongly on the part depth and total wall perimeter, with values from a few seconds to a

Fig. 2.2 ISF forming process and thinning phenomena

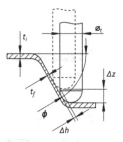

few hours. A reference value for the process time is commonly under 30 min for a 200 × 200 mm × 100 mm part, including the machine set up, blank clamping, forming and part release. The major setbacks of the ISF process are the surface quality and low accuracy. Part accuracy and surface quality depend strongly on the forming tool path and parameters, with typical better results when using small forming increments. The surface quality benefits from the lower friction possible. Oil lubrication is used and the tool is allowed to free spin or to spin to minimize friction. Different strategies can be used to improve accuracy. ISF can use over forming to compensate for the material spring back effect. Local heating can be used to both increase formability and accuracy.

> ISF key features are as follows:
> - Low cost, dieless operation;
> - Free-form capability;
> - Sheet metal forming by a progression of localized deformation;
> - Simple forming punch with CNC tool path control;
> - Compatible with different materials and thicknesses;
> - A wide range of achievable part dimension and design features; and
> - Fair part quality regarding accuracy and surface finishing.

2.2.1 Forming Mechanics

Incremental forming is a continuous process where the material is shaped by localized deformation. This huge difference when comparing to the conventional forming processes make the theoretical foundations of conventional processes not valid for ISF.

Another major difference between the conventional processes and ISF is related to the type of strain and the formability limit. The formability of incremental forming is higher than that of conventional forming processes such as stamping. The forming limits can be characterized by the maximum wall angle before failure occurs.

This maximum angle is dependent on the material type, sheet thickness and process parameters such as tool radius, step down, feed rate, local temperature of the sheet. Increasing sheet thickness, decreasing tool size and decreasing vertical step size all tend to increase formability.

The deformation in ISF process is a combination of shearing, stretching and bending. As the wall angle increases, stretching plays a more important role in deformation than shearing. When the wall angle decreases, the deformation is mainly due to shearing. The direct strain perpendicular to the forming direction has a major value. Biaxial stretching occurs at inner corners and plain strain stretching at the flatter sides, resulting in failure to occur mostly at the corners. The area near the clamped edge of the parts is the most affected by bending [7–9].

During the forming operation, the through-thickness strain is significantly affected, leading to a thinning phenomenon. The thickness distribution on a part, formed by ISF, can be approximated by the sine law, Eq. 2.1, where t_i is the blank thickness, ϕ is the wall angle and t_f is the final wall thickness as represented in Fig. 2.2. This estimation is effective for areas away from the bending affected region, as this has a smaller thickness reduction. High wall angles close to the maximum allowable value for a given material lead to greater thickness reduction due to the formation of microcracks [9].

$$t_f = t_i \times \sin(90° - \phi) \tag{2.1}$$

The formability limit of ISF operations can be represented by a forming limit curve (FLC) in a forming limit diagram (FLD) defined by a straight line with a negative slope. Figure 2.3 represents the formability limit configuration on an FLD where the slope of the FLC_{ISF} line is given by Eq. 2.2, usually varying from -0.7 to -1.3. While the FLC is determined concerning local necking failure, the FLC_{ISF} is the outcome from fracture. This is because necking phenomenon is suppressed in ISF. As the plastic deformation only occurs in a small zone surrounded by rigid material or at least only deformed elastically, a potential neck is unable to grow. In this way, it can be regarded that ISF is limited by fracture only with uniform thinning until fracture [2, 10].

Fig. 2.3 ISF forming limit curve [10]

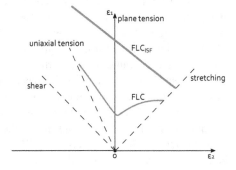

$$\frac{\varepsilon_1}{\varepsilon_2} = -\frac{5 \cdot \frac{\varnothing_t}{2 \cdot t_0} + 2}{3 \cdot \frac{\varnothing_t}{2 \cdot t_0} + 6} \tag{2.2}$$

2.2.1.1   Forming Forces

The forming force in ISF is very small in comparison to the deep drawing or stamping processes. Besides, the force intensity does not depend on the part size. That is why the production of very large products is absolutely appropriate for ISF. Major studies on the ISF forming forces have been done by experimental and numerical methods. The force can be decomposed in a vertical (F_z) and horizontal (F_{xy}) components. The horizontal component can be decomposed in a tangential (F_t) and perpendicular (F_r) component. Experimentally, it has been determined that the force, namely, its vertical component, is influenced by sheet thickness and material properties, tool diameter, wall angle and vertical step size or scallop height. The forming force increases when forming stiffer materials and with the increase of the sheet thickness. The forming force also increases with bigger forming steps and using bigger tools. Lubrication and tool rotation have no influence on F_z, affecting only the horizontal component of the force [11, 12].

The forming force increases when forming stiffer materials and with the increase of the sheet thickness. Besides, it appears that there are linear relations between the logarithms of the force and the sheet thickness. There is also an increase in the forming force with the increase of the draw angle. However, F_t is not influenced by the wall angle and F_z has a maximum value differs depending on the material. F_z is proportional to the function $\alpha cos\alpha$; physically, it means that the F_z is proportional to the draw angle and to the resulting wall thickness as approximated by the sine law. F_r also increases with the angle, being negative for small angles. In what concerns the tool diameter, there is also an increase in the forming forces components F_z and F_r when using larger tools. F_t in virtually unaffected. The forming force also increases when using larger vertical steps. A generalized formula which allows to predict the axial force F_z for any material based on the tensile strength only has been developed [11, 12].

Equation 2.3 has been experimentally determined to predict the generalized steady-state value of vertical forming force, where σ_u is the ultimate tensile strength and Δh is the scallop height. Some peak values of the vertical forming force are achieved, mainly when forming higher wall angles, thicker sheets and using bigger tools. A reference value for the vertical forming force can be obtained by the second member of Eq. 2.3. Equation 2.5 has been determined to approximate the radial forming force, where c depends on the forming material (e.g. 2.54 for aluminium alloys). The scallop height relates to the vertical step down according to Eq. 2.4. All the equation use linear dimensions in millimetre, angles in degree, stress in megaPascal and determine forces in Newton [12]. Equation 2.6 has been estimated to relate the tangential force to the radial force.

$$F_z = 0.0716.\sigma_u.t_0^{1.57}.\varnothing_t^{0.41}.\Delta h^{0.09}.\alpha.cos(\alpha), \qquad F_{z_{ref}} \approx 3.8.\sigma_u \qquad (2.3)$$

$$\Delta z = 2.\sin(\alpha).\sqrt{\Delta h.(\varnothing_t - \Delta h)} \qquad (2.4)$$

$$F_r = F_z.\tan\left(\frac{\alpha + \beta - 17.2°.(\varnothing_t/10)^{-c}}{2}\right), \qquad \beta = \arccos(1 - 2.\Delta h/\varnothing_t) \quad (2.5)$$

$$F_t \approx F_r.\left(360.\alpha^{-1.23}.\varnothing_t^{-0.62}\right) \qquad (2.6)$$

It is reasonably assumed that the rate of plastic work due to the tangential force acting on the forming tool is the product of the tangential forming force and the horizontal component of the feed rate speed [8]. Analogously, one can assume that the plastic work due to the tangential force is the product of the vertical forming force and the vertical component of the feed rate speed and the perpendicular component of the forming force produces no work. Thus, the forming power can be calculated by Eq. 2.7, where v_t is the tangential component of the feed rate speed and v_z is the vertical component of the feed rate speed:

$$\dot{W}_{forming} = \mathbf{F}.\mathbf{v} = F_t \times v_t + F_z \times v_z \qquad (2.7)$$

2.2.2 Process Strategies and Tool Paths

Whereby ISF works incrementally, where the tool follows a preprogramed path, the definition of the process strategies strongly influences the result. Simplest ISF operations are fulfilled in a single-stage three-axis strategy. More complex forms are accomplished with multistage and up to five-axis approaches. The selection of a forming strategy and tool path type is determined by the part geometry and influenced by the seek part quality and manufacturing time.

Most operations performed in ISF start the process from the part boundary and progressively deform the sheet through trajectories drawn. The tool tip follows the contours of the geometry, increasing the forming depth along the way. This increase is performed according to a defined vertical step size (Δz) with typical values from 0.1 to 2 mm. Depending on the slope variation, the part may benefit from using a constant vertical step or a constant scallop height (Δh). While constant Δz are usually faster, the use of constant Δh benefits the surface finishing mainly on parts with variable slope angle. In this approach, the tool follows a series of consecutive contours with variable step depth in order to keep constant the distance between passages.

Either when using constant Δz or Δh tool paths, generally, smaller forming increments lead to a better forming result. However, the use of small steps leads to longer manufacturing time.

The most relevant tool paths used in ISF are represented in Fig. 2.4. While the tool follows the outer contour of the piece at each forming step for every path type,

Fig. 2.4 Common tool path
strategies in ISF operations

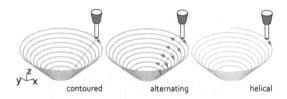

the transition between levels differs. In both contoured or alternating types paths,
the tool moves from one level on the z axis to the next diagonally in the xyz axes.
The difference between the strategies is a change in the movement direction between
levels which may lead to better results as it reduces part twisting. Although these tool
paths are easily defined in most CAM software, the step down movement creates a
line where the tool moved from level to level. As an alternative, a helical or spiral
type tool path can be used with tool head following a continuous path outlining the
shape of the geometry. The tool is always moving in the x, y and z axes, creating a
helix pattern in the formed part. Besides resulting in better surface quality, this tool
path type leads to better geometrical accuracy [13, 14].

> ISF forming operations use mostly contouring tool path strategies with vertical
> increments defined by a constant steps height or constant scallop height. The
> forming layer transition is either performed locally or with a shifted plunge
> position. Helical tool paths allow a continuous forming operation leading to
> better results mainly in what regards consistency along all surface.
> The major limitation of the single-stage strategies is the limit wall angle.
> Besides, a single-stage operation implies the use of a single forming tool,
> with further limitations in the small detail definition. Still, these tool paths are
> suitable for much parts and states an important aspect as the basis for more
> complex multistage operations.

The use of the different types of tool path presented is widely spread as they
are simple to generate using commercial CAD/CAM software and produce very
acceptable results in a time effective operation. However, this single-stage strategy
demands a geometrical complexity limitation. In order to expand the flexibility of
the ISF processes, multistage forming can be used.

There are two major mottos for the use of a multistage forming. On the one hand,
this strategy can be used to help improve formability, allowing greater draw angles.
On the other and, multistage can be used to improve accuracy and detailing in some
complex shaped parts.

On the first, the main goal is to enable the fabrication of parts with draw angles
above the single-stage forming limit. With this multistage strategy, it is possible to
fabricate parts with vertical walls and even with a negative draft. Figure 2.5 represents
a schematic of possible multistages approach to increase the draw angle to high

Fig. 2.5 Multistage
strategies in ISF operations

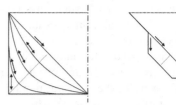

slope or vertical wall on the left. The strategy extends deformation to all the material available in the blank, seeking to use the maximum amount of material in the formed walls, minimizing thinning. The first stage stretches the blank into a 45 cone and the following stages gradually move the middle of this section towards the corner. All stages except the first can be performed with either downwards or upwards moving tool. Different combinations may be used to better adapt to a given geometry and depending on the target angle and seek part quality [15].

On the second case to use multistage, the goal is to seek for better accuracy and greater detail. This approach works in the same way as a finishing milling operation does, with the possibility of using different tools for forming different features of the part, as represented on the right of Fig. 2.5. This strategy is often used isolating the parts geometric features and improve the detail by forming one feature of the part at a time.

The presented option defines the tool path of the spherical tool tip. Most machines operate in three axes only, so that the tool tip follows the defined tool path with its axis aligned to z direction. As an alternative, five-axis strategies can be used to perform ISF operations. This solution becomes attractive due to its high flexibility, allowing the most convenient punch positioning at the surface sheet metal. At one point, the possibility to align to tool with the forming force allows using thinner tools, as bending will be minimized. This allows forming parts with greater detail, for example, smaller radii. Besides, the tool inclination promote lower friction coefficients and improve formability, allowing achieving steeper wall angles [16].

> The use of multistage forming operation allows to increase the feasible wall and the minimum part detail. On the other hand, these strategies lead to more time-consuming forming operation. The use of more complex five-axis tool paths may allow similar benefits in a more time efficient process.

2.2.2.1 Improving Part Quality

The use of simple ISF forming strategies allows achieving fair results. Still, ISF processes are characterized by a low accuracy and a rough surface finishing. Typical accuracy of the single stage formed parts is 2–4 mm, depending on part geometry and size. The surface finishing has typical Ra values up to 0.5 and Rz up to 3. Nevertheless, several techniques may be used to improve the part quality, bettering the process performance.

When the goal of a forming operation is the manufacture of a unique part, several strategies can be used to improve the part accuracy. As the inaccuracy is greater at lower slope walls, the right positioning of the part may benefit the part accuracy, at the cost of greater thinning. The use of multistage can also be used to improve accuracy, repeating final forming stages to correct the major deviations. The addition of stiffness geometric elements to large flat areas in the part design also improves the accuracy by minimizing the tent effect.

When the ISF operation is intended for a small batch, or even for a single part prototyping with the cost of a sacrificial part, different strategies may be used. By measuring a first ISF prototype, it is possible to identify the deviations from the desired surface. By mirroring the obtained part through the target contour one can obtain a new surface. This virtual surface can be used to generate a new tool path and reproduce the part at better accuracy [4].

The part accuracy is also influenced by the ISF process used. DPIF and TPIF allow achieving better results than SPIF, at the cost of higher initial cost in dedicated tooling manufacturing or highly complex hardware and tool path generation. On SPIF, the use of a well-adjusted backing plate has a large benefit on the part accuracy, particularly on the edge bending. The need for the dedicated tool may be avoided by forming the part inside a stabilization container to move away the part from the most inaccurate top edge.

Apart from benefiting the accuracy, other strategies may be used to improve the surface quality. The surface quality is affected by both the scratch marks and skinning effect due to tool contact. The right lubrication and the right selection of the forming parameters, particularly the vertical step, have a strong influence on the surface roughness. Both mineral oils, synthetic oils, bio lubricants and other synthetic liquids may be used as a lubricant. The greater the hardness of the material to form, the lower the necessary oil viscosity. An alternative strategy to improve surface quality implies the use of a sacrificial sheet material to protect the formed part. This strategy avoids the direct tool contact to the sheet metal part, eliminating the source of increased roughness.

A final strategy not only to improve the part quality but also to increase the formability involves the use of a heat source. The goal is to perform localized heating ahead of the forming tool. Depending on the material to be formed, the heat source may be a heat gun or a laser. Although this strategy may increase the achievable part complexity and benefit quality, it requires additional hardware to be used.

2.3 Machines and Tools

Depending on the ISF process to be used, there are differences in the required hardware. SPIF and TPIF with a negative die require the simpler hardware as they only require the motion of a simple tool, either in three or five axes. TPIF with a positive die requires additional axis to perform the blank vertical motion. DPIF require the use of two independent tool controllers with three of five degrees of freedom each. Simplified operations of DPIF-L may be performed with a single tool controller with both the forming tool and the support tool connected by a rigid frame.

There is no effective restriction on the type of machine that should be used for SPIF operations. In theory, any machine capable of performing position control in three or more axis could be used for forming operation. However, the strength needed for forming operation restricts and influences the machine architecture. Several machines have been developed or adapted with the goal of performing SPIF operations [17].

Thus, ISF operations can be done in both CNC machines, with industrial robots and in purpose build machines. The use of CNC machines and industrial robots has the advantage of a well-settled controller capable of precise motion control. The use of these types of hardware requires little adaptation to perform ISF operations, where the most relevant changes in the machines are the addition of a sheet clamping and backing plate systems, and the reinforcement of the machines. However, the use of these types of hardware finds a lack of strength and stiffness which limits the feasible material and thickness. Thus, CNC machines and industrial robots have been used to perform forming operations, mainly on thin sheets or low strength materials due to the loading capability limits. In order to overcome the strength, purpose-built machines have been developed. The use of height strength structures or parallel kinematics motion controllers allows developing machines capable of handling thick sheets and stiff materials as well as operating on five-axis tool paths.

Analogously, DPIF can also be performed using industrial robots or purpose-built machines. Most operations are performed with anthropomorphic robots and assume the designation roboforming. However, despite the great flexibility, this hardware configuration deals with significant limitations in what regards strength. The use of parallel kinematics bypasses these restrictions at the cost of minor flexibility.

A relevant purpose build ISF machine has been developed by Amino [18] and has gained notoriety for being the first company providing an incremental forming machine to the market. The machine uses a dual-point incremental forming (DPIF) process, requiring the use of a simple die for each part.

An important aspect of an ISF machine is the frame used for both supporting and clamping the blank. The most common approach is the use of a metal frame bolted all around the blank or the use of a toggle clamps-based system.

In most machines used for ISF operation, the forming tool is mounted on a standard tool holder or a custom design tool holder. For most applications, the tool holder is mounted on a spindle. The SPIF operation can either be done by allowing free passive rotation, minimizing friction or imposing a given spindle speed.

For most ISF applications, the forming tool or forming punch is a rod with a spherical end, mostly ball tip punches with typical diameter from 5 to 20 mm. Flat-ended tools with small corner radius can also be used for forming operations. The principal parameter of an ISF tool is its diameter and length. Furthermore, several tool shapes exist, particularly straight tools, tapered tools, dapper punches and tipped tools.

Being under high compressive and shear stress, the tool must avoid excessive displacement of the tip due to elastic bending and excessive deformation of the hemispherical surface. Hardened steel (55 HRC) is usually used and the tool diameter and shape is chosen according to the forming forces. Tapered and tipped punch allow the use of small radii notwithstanding the stiffness.

2.4 Part Design and Applications

Since the ISF process is new when compared to conventional forming processes, not many industrial applications have been developed yet. However, with the consistent results that it is now possible to obtain, the process dissemination around the industrial companies is starting. Thanks to the dieless operation, ISF allow the manufacture of sheet metal parts with minimal initial investment. Thus, incremental forming applications differ from conventional forming processes. Besides the applications in the prototyping field, SPIF can also be used for the manufacture of unique parts and small batches. This capability leads to a new business possibility, enabling the development of exclusive or custom products.

The fundamental description of the ISF process states a free-form capability. Nevertheless, as in all manufacturing processes, the forming operation must be performed within certain limits to ensure successful results. The selection of the most appropriate forming parameters and tool path strategy helps to minimize the risk of failure. Still, the manufacturing results are highly dependent on the part design. Therefore, the adequate definition of the part configuration and geometry is essential to reach a satisfactory part quality.

Following particular design guidelines during a part conception help to ensure its feasibility. With these orientations as a reference, the ISF processes are suitable for the manufacture of unique parts or low volume series. In general, ISF is cost-effective up to 300–600 pieces, depending on part size and complexity. From that level onwards, investing in dies becomes feasible. The process enables the production of parts of any size, where the dimension is only limited by available hardware. In comparison with conventional processes, the break-even point has a slight dependence on the size of the part produced. For relatively large parts there is an increase in the break-even point as compared to smaller part in ISF, thus increasing the reasonable cost-effective maximum number of parts.

In what concerns the typical industrial applications, the better suited identified fields are the manufacturing of prototypes, the manufacture of design and architecture components, the fabrication of spares for obsolete parts and the fabrication of components for small volume industries such as aeronautics. Other identified application fields is the customization of parts, making them unique for increased value products. The fabrication of custom-made products of various shapes and dimensions is also a great potentiality of the ISF processes. Lastly, the ISF processes have a potential industrial applicability in complementing other manufacturing processes to achieve hard to form features [19].

It is known that the product lifetime is highly dependent on the customers' request. While mass production is used for most of the industrial products, the demand for custom products is increasing. The concept of mass customization introduces the possibility to create individually customized products, with mass production volume, cost and efficiency. The use of flexible manufacturing systems (FMSs) allows converting individually customized designs into individually customized products, with mass customization volume, cost, efficiency, accuracy and reliability [20]. ISF systems are strong candidates to integrate FMSs for the manufacture of sheet metal parts.

2.4.1 Part Configuration

As mentioned, ISF processes form a blank sheet to the desired shape. Depending on the blank size and shape, and on the movement restrains, several part confirmations are possible to achieve. The configuration affects both the design and the processing of the part, influencing both the suitable sheet thickness, geometrical features and process tool path and parameters. Figure 2.6 illustrates the most relevant possible part configurations with one-quarter cuts. Each configuration is further described.

In addition to the shape configuration used on the forming operation, the final part results from further trimming and finishing operations. Thus, the end piece may appear different from any of the presented shapes (e.g. trimming operation can completely remove the flat area of the blank, succeeding on a full curved part).

Single Curvature Container Type Parts

Most parts manufactured by ISF consist of a single or multiple container type shapes. The part geometry comes out of a flat plate with a wall angle between 0 and ϕ_{max} from the edge to the centre. Inside the container, the part shape is virtually free-form. The container type part configuration is the most common configuration of ISF parts as it allows the most versatile part design possible. A container shape can either define a full desired shape or can be the result of the surface extension up to the flat blank plane. The container type part configuration leads to the most cost-effective compromise between part quality and process time in most instances.

Fig. 2.6 Possible part configuration

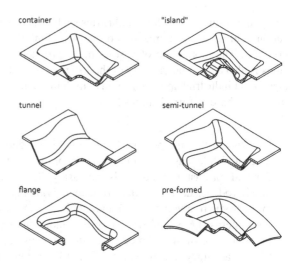

'Island' Type Parts

Despite the recommendation to design parts without changes in curvature, the ISF processes allow attaining higher complex parts, with two or more shifts in curvature. Nevertheless, depending on the used process, each change in curvature direction may require a flip of the blank and continue of the forming process in a new set-up. Simpler island configurations may only decrease the wall slope and be formed without a blank flip, but may lead to worst accuracy.

The ISF processes with multiple tools like TPIF or DPIF are better suited for this part configuration. The use of simple additional support to an SPIF procedure, in between SPIF and TPIF, can improve the decreased accuracy or avoid blank flips.

Tunnel and Semi-tunnel Parts

While the majority of the incrementally formed parts are formed with a fully constrained blank, it is possible to performed ISF while leaving one or two free sides. This way, it is possible to obtain tunnel and semi-tunnel type parts, maintaining the free-form capability in the part walls.

Tunnel and semi-tunnel parts are widely used in different products and applications. The use of tunnel or semi-tunnel part configuration benefits the useful area of the part, reducing the amount of scrap and allowing the fabrication of bigger parts. In large parts, this configuration also benefits process time as it dismisses the need for extended surfaces. A setback on forming tunnel or semi-tunnel like parts is the reduced maximum forming angle due to premature failure by tearing.

Flange Type Parts

Another option of ISF processes is to form free-form flanges along parts. Both inner and outer flanges can be formed. While inner flanges can be performed with freedom within all the part, outer flanged lead to higher limitation to be compatible with the blank clamping. As in the tunnel part configuration, ISF is properly suited for curve flanges, as other conventional processes are better suited for straight line flanges. The forming limits of the flanges are typically higher than in container type parts. However, the maximum wall angle and flange height are influenced not only by the forming parameters but also by the pre-cut hole shape.

Pre-formed Parts

A distinct option of ISF processes implies the use of other forming processes in an initial phase. ISF can be used to complement and surpass the limitations of other forming processes in the fabrication of prototypes, unique parts or small batches or for part customization of mass production frames.

A pre-formed part configuration replaces the flat blank by a stretch formed, deep drawing or bent part. Any type of the presented configurations can be used in a pre-formed part. Major restrictions result from an increased difficulty on clamping the part and higher complexity on the tool path generation. In addition, the higher stiffness achieved by some forming processes limits the possibility of performing ISF.

A derivative part configuration is the post-formed part. After initial operations of ISF process, the part can be finished by complementary forming operations like bending or coinage. Welding operations are also often used in post-forming stage.

Tracing Type Parts

A simplified version of a different configuration is the definition of tracing parts, where ISF is a suitable process to form marks and low reliefs in sheet parts. A single or a few passage tool path strategies can be used to simply locally trace a sheet instead of producing a deep 3D shape (e.g. adding texture or lettering). Tracing operation can either be accomplished in flat sheets, incrementally formed parts or pre-formed parts.

> A combination of multiple configurations in one part is possible and leads to greater design possibilities. Besides, the definition of the part configuration is frequently the result of the part surface extension to better adapt to the blank geometry. Exploring different possible orientations for a part manufacture can redefine a part configuration and often simplify the manufacturing design to more time and cost-effective process.

2.4.2 Design Guidelines

A good part design starts with the right selection of the material and thickness. It is important to note that the ISF parts are completed with other accessory processes as trimming, surface finishing, welding and others. Thus, the proposed guidelines for incremental forming should be considered attached to accessory processes restrictions. A good part project should look to the part feasibility and develop a detailed and accurate CAD model. The feasible limits of ISF have been purposed. Figure 2.7 summarises the most adequate part size and features, as well as the achievable part quality [21].

Materials and Thickness

Although ISF processes are better suited for high formability materials, the processes can be used with both metal and plastic sheet materials. Table 2.1 summarises some of the most used materials and thicknesses.

The selection of material and thickness should takes into account the required mechanical behaviour for the formed part as well as the consequent restrictions regarding the limitations of geometric features. In what concerns the mechanical behaviour of the formed part, thinning must be considered.

Geometric Features

The most important feature when designing an ISF part is the wall angle ϕ. Furthermore, as specified by Adams and Jeswiet [22] or Afonso et al. [23], limiting the part size to the available area and minimizing the number of curvature variations are also relevant.

Fig. 2.7 Achievable mechanical features and part quality through ISF [21]

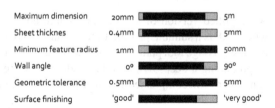

Table 2.1 Materials and thicknesses suitable for ISF

Material	Alloy	Thickness (mm)
Aluminium	1xxx	0.5–3.0
	3xxx	0.5–3.0
	5xxx	0.5–3.0
Steel	low carbon	0.5–2.0
	DPxxxx	0.5–1.5
Brass	—	0.5–1.5
Cooper	—	0.5–1.5
Titanium	Grade 2	0.5–1.5
PE	—	1.0–5.0
PA	—	1.0–5.0

In addition to the defined orientation lines, generic extended guidelines for an ISF part design are proposed. Figure 2.8 illustrates some of the most relevant guidelines:

- Avoid exceeding the reference ϕ_{max} angle, available in literature and Table 2.2 for different materials and thicknesses (Fig. 2.8a). Better results occur if the slope angle is kept away from the maximum value.
- Larger slope angles are possible at the cost of larger forming time and higher process complexity and risk of failure using multistage strategies (Fig. 2.8b).
- Undercuts are possible at even more difficult and failure risk.
- Minimize changes in curvature between convex and concave shapes (islands). Smooth curvature changes that don't create islands are possible (Fig. 2.8c and d).
- Avoid the definition of peninsula type geometries, mainly when the part is to be formed without a backing plate.
- Avoid abrupt gains of forming angle in a sloped wall.
- Parts may include more than one design feature, combining different shape configurations (e.g. multiple containers or combination of tunnels and containers).
- Limit the part area and depth to the machine workspace.
- Adequate the part size to the available sheet clamping system.
- Avoid large flat sloped areas, unless accuracy for that region is not an issue.
- Avoid shallow sloped walls, unless accuracy is not an issue.

Fig. 2.8 ISF part design guidelines

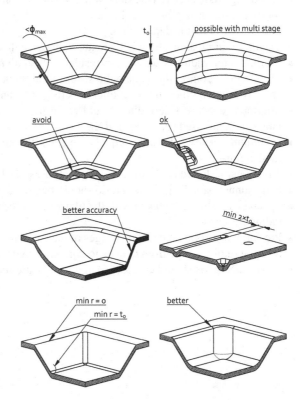

Table 2.2 Maximum reference wall angle for single passage ISF of container type parts [24]

Alloy	Thickness (mm)	ϕ_{max} angle
AA 1050-O	1.5	76°
AA 3003-O	0.85	70°
AA 3003-O	1.2	71°
AA 3003-O	2.1	78°
DP600 steel	1.0	68°
DP780 steel	1.0	42°
DP1000 steel	1.0	39°
HSS	1.0	65°
Brass	1.0	40°
Cooper	1.0	65°
Polyethylene	3.0	81°

Smaller angle for tunnel type parts

Larger angle for flange type parts

- Add reinforcement indents to large areas to improve accuracy and stiffness.
- Avoid large radii between the slope walls and the bottom or the top, unless accuracy for that region is not an issue (Fig. 2.8e).
- Evenly distribute the depth of the part.
- The minimum feature should be greater than twice the sheet thickness (Fig. 2.8f).
- Minimum top radius is zero, better results with small radius (Fig. 2.8g).
- Minimum bottom radius equals the sheet initial thickness, better results using a value two to ten times the sheet thickness (Fig. 2.8g and h).
- Minimum concave side radius should equal the bottom radius, better results if larger radii are used (Fig. 2.8g and h).
- Minimum convex side radius equals the sheet thickness, better results if larger radii are used.
- Tunnel parts should be narrow, wide tunnels are possible at higher risk of failure.
- Semi-tunnel parts length/width ratio should be greater than 1.
- Special care should be taken with the blank cut to form flanges, mainly on small radius corners.
- Pre-formed parts should avoid changes of the curvature direction, mainly on small bending radii (change from convex to concave or vice verse).
- Pre-formed parts should consider features to ensure fixation and reference.

Following the purposed guidelines helps to ensure it is possible to form a designed part without tearing or any other major failure. However, experimental testing may be required for some complex designs. Further, the desired accuracy or surface finishing required for the part may influence or limit the choice of the process.

It is possible to work around some guidelines at the cost of multistage, multitool and multi-set-up strategies. Nevertheless, these possibilities lead to higher failure risk and are much more demanding on a technological level. While in prototyping application it might be acceptable to increase the forming strategy difficulty, the fabrication of final parts, either unique or small batches should be more cautious.

It is also important to keep in mind that thinning occurs during ISF and it affects the final shape. When the design requires the use of the bottom surface, the final thickness should be considered. If the part has structural functions, it is recommended to analyse the mechanical behaviour of the formed part.

2.5 ISF as a Rapid Tooling Process

In addition to the manufacture of unique and low batch sheet metal parts, ISF as a potential field of application as a rapid tooling process. The RT concept is overall defined as a combination of processes to produce tools and moulds for conventional manufacturing processes in less time and at a lower cost relative to traditional machining methods. Being compatible with FMSs, with parts manufactured from CAD models without considerable dedicated tools in short time, ISF processes can be seen analogously to AM technologies and considered rapid smart manufacturing processes.

The use of sheet metal and other thin-walled moulds has considerable application in industrial processes, particularly in processing of thermoplastics and composite materials. These tools have usually attractive weight/strength ratios with low material costs and, when applicable, low thermal inertia. However, conventional manufacturing processes have limited geometric freedom or are very time-consuming and expensive. Furthermore, traditional RT techniques, using AM processes, struggle to achieve thin geometric features, becoming weak competitors for the mould manufacture.

The ISF processes have a great pretence for the fabrication of these geometries. Thus, ISF processes are highly suitable for direct hard tooling. In such a way, different ISF techniques can be used for the development of sheet metal moulds for various manufacturing process.

ISF may be used mainly as a direct hard tooling rapid tooling process. The manufactured parts, either from aluminium alloys, steel or other materials, can be used in multiple cycles. Still, it is important to consider a correct sizing of the parts strength and stiffness in both a mechanical and thermal way. It is also important to refer that often the ISF technologies must coexist with others in order to achieve the desired geometry and properties.

References

1. S. Tabibian, M. Najafabadi, Review on various kinds of dieless forming methods. Int. J. Eng. Adv. Technol. (IJEAT) 3(6), 24–28 (2014)
2. M. Skjødt, N. Bay, T. Lenau, Rapid prototyping by single point incremental forming of sheet metal. Ph.D. thesis, Technical University of Denmark (2008)
3. M. Tisza, General overview of sheet incremental forming. J. Achiev. Mater. Manuf. Eng. 55(1), 113–120 (2012)
4. J. Jeswiet, F. Micari, G. Hirt, A. Bramley, J. Duflou, J. Allwood, Asymmetric single point incremental forming of sheet metal. CIRP Ann. Manuf. Technol. 54(2), 88–114 (2005)
5. J. Jeswiet, D. Adams, M. Doolan, T. McAnulty, P. Gupta, Single point and asymmetric incremental forming. Adv. Manuf. 3(4), 253–262 (2015)
6. E. Leszak, Apparatus and process for incremental dieless forming, US Patent, US3342051A1 (1967)
7. D. Malwad, V. Nandedkar, Deformation mechanism analysis of single point incremental sheet metal forming. Procedia Mater. Sci. 6(2014), 1505–1510 (2014)
8. Y. Li, W. Daniel, Z. Liu, H. Lu, P. Meehan, Deformation mechanics and efficient force prediction in single point incremental forming. J. Mater. Process. Technol. 221, 100–111 (2015)
9. G. Hussain, L. Gao, A novel method to test the thinning limits of sheet metals in negative incremental forming. Int. J. Mach. Tools Manuf. 47(3–4), 419–435 (2007)
10. M. Silva, M. Skjoedt, A. Atkins, N. Bay, P. Martins, Single point incremental forming & formability/failure diagrams. J. Strain Anal. Eng. Des. 43(1), 15–36 (2008)
11. A. Petek, K. Kuzman, J. Kopac, Deformations and forces analysis of single point incremental sheet metal forming. Achieves Mater. Sci. Eng. 35(2), 107–116 (2009)
12. R. Aerens, P. Eyckens, A. Van Bael, J. Duflou, Force prediction for single point incremental forming deduced from experimental and FEM observations. Int. J. Adv. Manuf. Technol. 9(12), 969–982 (2009)
13. A. Fischer, Embodied computation: exploring roboforming for the mass-customization of architectural components. Masters thesis, Carnegie Mellon University, 2015
14. M. Skjødt, M. Hancock, N. Bay, Creating helical tool paths for single point incremental forming. Key Eng. Mater. 344, 583–590 (2007)
15. M. Skjødt, M. Silva, P. Martins, N. Bay, Creating helical tool paths for single point incremental forming. Key Eng. Mater. 344, 583–590 (2007)
16. J. Sá de Farias, R. Bastos, J. Ferreira, R. Alves de Sousa, Assessing 3 and 5 degrees of freedom tool path strategy influence on single point incremental forming. Key Eng. Mater. 651–65, 1159–1162 (2015)
17. S. Marabuto, D. Afonso, J. Ferreira, F. Melo, M. Martins, R. Alves de Sousa, Finding the best machine for SPIF operations-a brief discussion. Key Eng. Mater. 473, 861–868 (2011)
18. S. Aoyama, H. Amino, Y. Lu, S. Matsubara, Apparatus for dieless forming plate materials, Europaisches Patent EP0970764 (2000)
19. J. Sá de Farias, J. Ferreira, S. Marabuto, A. Campos, M. Martins, R. Alves de Sousa, Towards smart manufacturing techniques using incremental sheet forming, in *Smart Manufacturing Innovation and Transformation: Interconnection and Intelligence* (IGI Global, Hong Kong, 2014)
20. S. Smith, G. Smith, R. Jiao, C. Chu, Mass customization in the product life cycle. J. Intell. Manuf. 24(5), 877–885 (2013)

21. K. Jackson, *Incremental Sheet Forming* (Institute for Manufacturing, University of Cambridge, 2005)
22. D. Adams, J. Jeswiet, Design rules and applications of single-point incremental forming. Proc. Inst. Mech. Eng. Part B J. Eng. Manuf. **229**(5), 754–760 (2014)
23. D. Afonso, R. Alves de Sousa, R. Torcato, Integration of design rules and process modelling within SPIF technology-a review on the industrial dissemination of single point incremental forming. Int. J. Adv. Manuf. Technol. **94**, 4387–4399 (2018)
24. A. Behera, R. Alves de Sousa, G. Ingarao, V. Oleksik, Single point incremental forming: an assessment of the progress and technology trends from 2005 to 2015. J. Manuf. Process. **27**, 37–62 (2017)

Chapter 3
Complementary Manufacturing Processes

3.1 Sheet Metal Forming Processes

A large variety of conventional sheet metal forming processes are globally available. While some require dedicated tools for each new design, others are considered dieless or use multipurpose tools. These forming processes are often used to complement the manufacture of incremental forming parts, both before and after the automated process. The use of complementary forming processes finds its applicability not only to support or increase the geometry freedom of ISF parts but also for the manufacturing of simpler parts that complete the tool assembly.

Apart from the presented incremental forming methods, dieless sheet metal forming processes with greater suitability to support rapid tooling can be grouped into three groups: bending, manual metalwork and spinning.

Bending

The simplest method of metal forming consists of bending a sheet in a straight line. Bending operations are mainly held using multiple proposed tools, drastically reducing the initial investment. However, the use of this type of tools strongly harm continuous operation they are labour and time demanding. Special propose tools may also be used but at a greater cost.

There are several techniques for straightforward bending, depending on part design and production volume. The most common are V-bending and U-bending, folding or wiping, roll bending, and roll forming. Figure 3.1 represents basic bending operation processes.

V-bending can be used to bend angles up to 180°, though bending to 90° is more usual. The process involves punch pressing a sheet down into a V-shaped die. The bent angle is controlled by the punch stroke, ending its movement on the air, and the bent radius depends on the width of the die and thickness of thin sheets. Spring back is very noticeable in V-bending operation. However, it can be compensated by the punch stroke control or minimized by bottoming. The process creates on straight bent

D. Afonso et al., *Incremental Forming as a Rapid
Tooling Process*, Manufacturing and Surface Engineering,
https://doi.org/10.1007/978-3-030-15360-1_3

Fig. 3.1 Bending processes

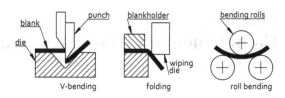

per knock with a restriction of a minimum angle length due to the V die. Two straight bents may be done at once using U-bending. To form parts with fewer restrictions in what concerns angle size folding or wiping can be used. In these processes, the sheet is clamped on one angle and the other is bent be either sliding a wiping die or turning a swing beam. As in V-bending, both angle and radius are influenced by sheet thickness and tools stroke and geometry. A minimum clamping length should also be fulfilled to allow flawless operation.

Both V-bending and wiping are restricted to small radius bents. To manufacture large continuous radius roll bending is better performed. This process uses rollers, three minimum, to form sheet into cylindrical or conical sections. The movement of roller axis position, changing the wheelbase distance, allows to control the forming radius. Several configurations of roller position can be used to allow working with different thicknesses and rolling the sheet to the leading and trailing edges, avoiding straight ends. In cone bending, breaking latches must be used to avoid or limit the sheet movement, allowing to roll both longer and shorter ends to the same angle [1, 2].

In bending operations, the plastic deformation only occurs in the bent region and the rest of the sheet is only affected by the consequent displacement. This area is under tensile elongation in the outer surface of the sheet and compressive strength in the inner side of the sheet. The tensile and compressive denting strains increase with smaller forming radius. When using small radius, major issues with bending operation involves cracking on the outside bend surface and thinning of the bent area. When using large radius, major difficulty is to control accuracy and repeatability as the process suffers appreciable spring back. If the sheet is relatively narrow, a contraction in the width may also occur [1].

As many operations principles are used for bending operations, each has specific design guidelines. Summarizing, attention should be paid to minimum angle length, distance between bents, withdrawal of cut-offs from bent lines, minimum radii and other features related to the blank cut.

Manual Sheet Metalwork

The bending process have great potential to complement the ISF for bents, both in pre-formed and post-formed operation. However, their use is limited to straight bents and planable shapes. When this is not the case, the use of hand-skilled processes has

Fig. 3.2 Manual work forming processes

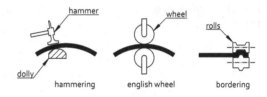

great potential both to define pre-formed blanks, to form details on ISF parts or to locally improve parts accuracy, particularly on the top edge.

One of the oldest processes in sheet forming is Hammering. This process was initially done manually using different shapes hammers, dollies and mallets. Since the beginning of the twentieth century the English wheel has been used as wheel to help curve and improve smoothness. Figure 3.2 represents manual hammering and English wheel operation. To improve productivity, power hammers, mechanical or pneumatic, have been used replacing the manual tools. There are two different principles of hammering. On the one hand, deformation is caused by the energy of the moving part and its impact. On the other hand, the predetermined movement directs the forming energy to a specific area [3].

The hammering techniques take advantage of the use of different shape of hammerheads and dollies. The manual sheet metal work is complemented with the use of a large variety of tools, namely, flat jaws, locking, round mandrel and universal pliers. These tools have primer applicability in shaping thin or low strength materials. Hammering major limitations are related to the operator skills. Besides the process is very time-consuming and repeatability is difficult to archive. Still, the use of these processes has great potential in both free-form geometries and small details in sheet metal parts.

In recent years, hammering has been taking advantage of motion control and uses a robotic arm to handle the hammer tool and form the sheet, which is clamped in a support frame. By moving a hammering tool over a sheet of metal fixed in a frame, a three-dimensional workpiece can be produced without using any special die plate. The use of a common industrial robot is possible because the forces involved in the process are considerable low [4].

An alternative manual forming process that can be used to complement ISF parts is sheet metal bordering. The process uses pleat rolls or flange rolls to shape the metal sheet along both straight and curved lines. Both manual crack and electrical machines can be used. Although the bordering machines could be used both for flanging operation and for mid-surface pleating, its operation as a complementary process to ISF finds its main applicability dealing with the parts edges.

Fig. 3.3 Metal spinning
processes

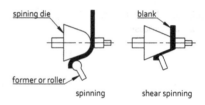

Spinning

Apart from the presented processes, other incremental forming processes like spinning or shear forming also finds an interesting relevance to complement the sheet metal tool making.

Although these processes require the use of a die, its simplicity allows a very fast and economic development. Thus, when parts include revolution bodies, these processes may be an appealing alternative to ISF.

Spinning is a very old method of producing axisymmetric parts. The process uses a lathe to turn a circular blank while a forming tool pushes the material to a spinning die in a series of sweeping strokes. Operation can either be performed by hand, by a copy or CNC process. Figure 3.3 represents forming spinning operations. Although the figure only presents external forming operation, internal option is also available. The final shape is acquired by gradually laying the material against the die. The blank has the same area as the final part so that the wall thickness remains more or less unchanged as is process is roughly an axisymmetric bent. The process allows a good surface finish. Despite low tooling costs, labour costs can be high unless operations are automated. Spinning operation can also be performed with the sheet held by outer frame, operation in the middle.

The spinning process involves low forces but there is a high risk of fracture due to excessive hardening and the parts have high residual stresses. As the blank must be rotated in a lathe, a size and mass limitation must be ensured.

The most common defects in spinning are wrinkling, circumferential cracks and radial cracks. Wrinkling occurs due to high compressive circumferential stresses buckling the flange. To avoid wrinkling, a combination of tensile and compressive stresses in the material needs to be introduced gradually. High tensile radial stress may cause circumferential cracks. Radial cracks may form in two different cases; due to circumferential tensile stresses or a combination of circumferential compressive and bending stresses which occur when existing wrinkles are being worked [5].

In the process of shear spinning or flow turning, also represented in Fig. 3.3, the strain is obtained by stretching the material over the die. The blank has thereabout the same diameter as the final shape bu thicker. The shape is archived as forming tool stretches the material over the die with high thickness reduction, according to the forming angle.

As the deformation only occurs in the point of contact between the forming tool and the sheet, the remaining material remains stress-free. This allows a greater degree of deformation to be archived then in other processes like spinning or drawing, as the flange remains virtually stress-free. Operation forces as much higher than in spinning precluding the hand working.

3.2   Cutting and Drilling

Sheet metal cutting is one of the most important requirements for the manufacturing of sheet metal tooling. Different cutting operations are required before the forming operation to create blanks and after forming to separate the useful section of the formed shape. A wide variety of cutting processes can be used, using both shearing, mechanical material removal and thermal processes. Figure 3.4 illustrates some of the most relevant processes, which are described below.

Sheet metal cutting applications are commonly associated with press operations. Either as the final step to separate the formed product from the blank or as an intermediate operation, the basic principle involves the separation of one defined area from the remaining sheet, which is attained through shearing. Forces caused by the upper blade edges and the lower blade ones promote the cutting process. Using presses, the cutting processes is referred to as punching. The upper and lower blades of the cutting tool are treated as the punch and die. The distance and angle between upper and lower tools can be adjusted according to the material used and the desired finish. Precision sheet cutting can be obtained by using compression tools around the cutting perimeter [2].

While most press shear cutting processes involve the use of dedicated punches and dies, thus not suitable for unique parts associated with rapid tooling. Nevertheless, shear cutting is mostly used in more versatile equipment to prepare blanks. The use of guillotines allows a straight cut which allows to create the most common blanks outer shapes. The same process can be used with thicker sheets to support the productions of backing plates.

A more flexible use of shear cutting is done using CNC punching, allowing to incrementally cut free-form shapes. A group of tools consisting of sets of punches and dies with small simple shapes are used to perform small cut in the sheet at designated places. The controlled repetition of those cuts allows to create a virtually free design blank.

Despite the cut cleanness and speed of the shear processes, they are mostly limited to flat sheets, thus to blank preparation. As an option, manual shearing tools like shear cutter drills, nibbler cutters or gauge shears can be used in free-form surfaces but limited to the human precision.

As an alternative, sawing may suit the simplest types of cutting processes, either straight cut-offs or cutting along a simple path, manually or CNC driven. The resultant piece may require additional finishing operation due to the sawing intrinsic process characteristics. Nevertheless, sawing can be used along a flat sheet of free-form surface, using either a bandsaw or rotational tools.

Fig. 3.4 Sheet metal cutting processes

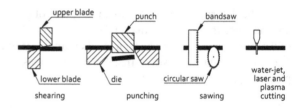

Fig. 3.5 Sheet metal drilling and tapping processes

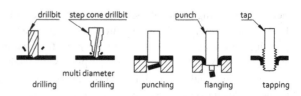

To promote better finishing and/or to cut harder, thicker materials (e.g. high strength steels), water-jet, laser or CNC plasma cutting technologies have been becoming a very popular solution. These technologies are easily automated using CNC control in three or five axes, being one of the most flexible cutting solutions. The three option differ in the working principle, with different working speeds and finishing results while adapting to different materials and thicknesses [6]. In addition, wire electrical discharge cutting technology can also be used. Most cutting equipments from these four technologies can be used for cutting backing plates and blanks on flat sheets. The use of higher complexity equipments allows performing cuts on 3D-shaped formed parts. Nevertheless, to perform these cuts in ISF parts it is often required to perform stress relief heat treatments to the parts before being unclamped.

A particular cutting process is commonly required in sheet metal parts for rapid tooling to open holes, either for fixing, guiding or venting. While some of the described processes allow the opening of holes, there are other technologies dedicated to this operation. Drilling refers to any cutting process that uses a drill bit to open-cut or enlarge a circular hole (some novel technologies may allow for square holes). The drill bit is a rotary cutting tool, which is pressed against the workpiece and rotates at rates from hundreds to thousands of RPMs, Fig. 3.5. The combination of tangential and compressive forces cuts off chips of material (swarf) from the hole promoting the drilling operation [7]. The process can either be achieved using hand tools or CNC-controlled motion.

While drilling is the most commonly used principle for opening holes in formed sheet metal parts, other processes can also be used. Punching, as mentioned above, is a particularly effective technology for opening holes in flat sheets. Apart from the controlled CNC equipments, manual hole punching tools can be used for this process. The use of punching techniques adds a greater possibility of opening non-circular holes. Additional, the use of knockout hole punching tools allows the operation in flat areas of non-flat parts, enabling to work on the walls or bottom of formed parts.

A process derived from punching takes advantage of the conformation capability of the material beyond the shear cutting. Hole-flanging is a process widely used in industry and with great applicability in ISF parts for rapid tooling. It involves creating or enlarging a hole via stretching, which stiffs the material and creates a flange. Practical applications include locating pilots, pre-holes for threading, and assembly locations.

The threading operation on flanged holes in formed sheet metal parts is majority done by tapping. Taps are cutting tools used to create screw threads, cutting the female portion of the mating pair. Tapping can be accomplished using automated tools, but in rapid tooling parts manual work is often preferred to achieve higher flexibility in a unique part. Moreover, while manual hole taping operations are usually done with three steps tools named tapper, plug and bottom, on flanged holes the use of tapper is commonly sufficient due to the short length and hole clearance.

3.3 Welding and Joining

As most rapid tooling designs often require the assembly of multiple ISF parts or sections, joining operations assume a fundamental role in the tool development. These joining operations can be grouped into permanent and non-permanent or demountable connections.

Permanent links between sheet metal sections are mostly used when the size or the complexity of a tool part do not allow the forming operation of a single piece. These permanent methods include welding, adhesive bonding or material deformation methods using rivets or the sheet metal material itself, as illustrated in the upper line of Fig. 3.6. Non-permanent links can be established either between two sheet metal parts or between one sheet metal part and a bulk material: metals, polymers or wood and derivatives. Most non-permanent connections use threaded components, with a large number of possibilities as illustrated in the second line of Fig. 3.6.

Welding assumes one of the most important roles for joining sheet metal parts. The most relevant welding operations are performed to join different formed parts into more complex geometry. A wide variety of processes and configurations can be used. Most common configuration for parts edge configurations in thin sheet metal tools are square edge butt, overlap and corner. Flange and tee may also be used, mainly for reinforcements or supports. Arc welding using TIG is one of the most compatible process with thin sheets, although it requires high skilled operation. Pulsed MIG can be a good alternative of a faster and easiest weld, although it requires more post finishing. Spot welding is even a fastest and easiest alternative, while limited to overlapping edge configuration, limiting its use in moulding surfaces [8].

Fig. 3.6 Sheet metal permanent and demountable joining processes

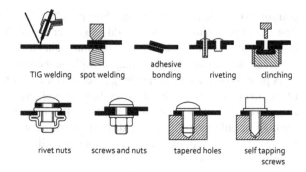

Another application for welding operation is to create tailored blanks, either using different materials or different thicknesses. The joining of dissimilar materials or thicknesses is rather difficult or impossible with conventional welding technologies. The use of friction stir welding can work around the limitation for the creation of tailored flat blanks [8, 9].

As an alternative to welding, adhesive bonding can be used to joint two parts with overlapping edges, being compatible with dissimilar materials and thicknesses, as well as compatible with free-formed shapes. The selection of the adhesive must be made taking into account the inaccuracy of the forming process, especially when joining free-form surfaces, and the joint must be designed considering the loading directions. Apart from joining multiple parts into a more complex geometry, adhesive bonding is often used to allow overlapping reinforcement plates [9, 10].

Material deformation methods can also be used for permanent connections, although not often compatible with the moulding surface of rapid sheet metal tools. Material deformation methods include the use of external deformable parts like rivets, self-piercing rivets, one-sided high-speed nails and crimps, or the single use of the material of the two sheets like in clinching and hemming [10, 11].

While most joining operations in rapid sheet metal tools are permanent, the need for demountable connection should not be neglected. The use of fasteners assumes the most relevant role in these situations. Depending on the tool design, these links can be established between sheet metal parts or between sheet metal and other materials. In the first case, the use of rivet nuts is one of the best options, facilitating assembly and disassembly operations. When using connections to other materials, the fastener selection is mostly dependent on the interface.

3.4 Finishing

The mechanical surface of a workpiece has been defined as the set of features in a workpiece which separates the entire workpiece to the surrounding medium. This surface quality, particularly the surface finishing, is crucial to ensure the part quality and performance, namely, in tool making where the surface defects will propagate to the result products. The development of methods to improve a surface quality, as well as measuring it, have great importance for a number of reasons: prevention of mechanical failures in parts related to surface defects; improvement of part contact in assemblies; reduce the risk of corrosion; possibility of performing surface treatment; component performance (e.g., light reflection and refraction, matter adhesion, moulded material ejection); aesthetics reasons [12].

As referred, ISF processes are characterized by a low accuracy and a rough surface finishing, with typical surface roughness values Ra values up to 0.5 and Rz up to 3. These values are commonly not compatible with tooling, thus, requiring some finishing operations. Two different approaches are possible to improve surface finishing in sheet metal parts formed by ISF, using subtractive or addictive techniques. Negative or subtractive surface finishing operations can lead to great results. On the other hand, these processes are often very time-consuming and dependent on skilled human labour. In addition, some subtractive techniques may not be compatible with very thin sheets or may struggle to deal with part inaccuracy and differences in material properties and original surface finishing along the parts, typical from ISF parts. This is most common when using automated polishing methods. Thus, positive or addictive surface operation can be used instead, not only benefiting surface finishing but accuracy as well. A better surface finishing and performance is usually achieved using a combination of subtractive and addictive processes.

Mechanical finishing operations used in subtractive methods typically includes deburring, grinding, polishing, buffing and final visual inspection of a workpiece [13]. A large variety of tools and methods can be used to achieve a smooth surface on a sheet metal part. These operations can be achieved using either hand tools or automated processes. Some of the most common include the use of angle grinders or the use of rotational spindles, as represented in the first two examples of Fig. 3.7, or using full manual methods. All theses methods can be performed using grinding stone wheels; axial or radial sand paper flap wheels; sanding paper sleeves or disks; abrasive belts; brush wheel; polishing or buffing wheels; and polishing berets and caps or foam pads.

Addictive surface finishing take advantage on coating or painting techniques, as represented in Fig. 3.7. Apart from being used for improving surface finishing itself, these processes add the advantages of oxidation control and enhanced corrosion resistance. Coating techniques compatible with sheet metal parts include powder spray coating, resin coating, physical or chemical vapour deposition, electrochemical deposition, plasma spraying, and solgel process. Thin film coatings can be used for surface finishing with the main goal of improving the surface performance (e.g., corrosion, wear resistance) [14]. Thick film coating add the possibility of overlapping

Fig. 3.7 Sheet metal
finishing processes

grinding   milling   coating

original surface defects and deal with part inaccuracy, particularly in over-formed areas. This is mostly possible with resin coating operations where the addition of bulk volumes of material is possible. Nevertheless, the tool operating conditions, particularly thermal, must be taken into account for the choice of adding material.

A recent alternative for addictive surface finishing of ISF parts has been proposed by creating polymer–metal–composites. The sheet metal part is covered using a reactive adhesive agent that adheres to an over-moulded harder polymer. The adhesive between the sheet metal and the outer polymer acts as a stress and strain compensation layer, caused by thermal expansion and external loading, thus being a relevant option for rapid tooling [15].

References

1. J. Schey, *Introduction to Manufacturing Processes*, 3rd edn. (McGraw-Hill, Michigan, 2000)
2. H. Lundh, P. Bustad, B. Carlsson, G. Engberg, L. Gustafsson, R. Lidgren, *Sheet Steel Forming Handbook—Size shearing and plastic forming*, 1st edn. (SSAB Tunnplat AB, Göteborg, 1998)
3. J. Downie, *Power Hammer Techniques and Applications for Creating Compound Curves in Sheet Metal* (International Specialised Skills Institute, Melbourne, 2010)
4. T. Schafer, R. Schraft, Incremental sheet metal forming by industrial robots. Rapid Prototyp. J. **11**(5), 278–286 (2005)
5. O. Music, J. Allwood, K. Kawai, A review of the mechanics of metal spinning. J. Mater. Process. Technol. **210**(1), 3–23 (2010)
6. D. Krajcarz, Comparison metal water jet cutting with laser and plasma cutting. Procedia Eng. **69**, 838–843 (2014)
7. J. Walker, *Machining Fundamentals* (The Goodheart-Willcox Company Inc., 2004)
8. P. Kah, R. Suoranta, J. Martikainen, Joining of sheet metals using different welding processes, in *Proceedings of 16th International Conference* (Mechanika, Lithuania, 2011)
9. T. Sakiyama, G. Murayama, Y. Naito, K. Saita, Y. Miyazaki, H. Oikawa, T. Nos, Dissimilar metal joining technologies for steel sheet and aluminum alloy sheet in auto body. Nippon Steel Tech. Rep. **103** (2013)
10. S. Modi, M. Stevens, M. Chess, *Mixed Material Joining Advancements and Challenges* (Center for Automotive Research, Ann Arbor, Michigan, 2017)
11. K. Mori, N. Bay, L. Fratini, F. Micari, A. Tekkaya, Joining by plastic deformation. CIRP Ann. **62**(2), 25–28 (2013)
12. L. Blunt, The history and current state of 3D surface characterisation, in *Advanced Techniques for Assessment Surface Topography* (2003)

13. E. Kalt, R. Monfared, M. Jackson, Towards an automated polishing system—capturing manual polishing operations. Int. J. Res. Eng. Technol. **5**(7), 182–192 (2016)
14. P. Martin, *Handbook of Deposition Technologies for Films and Coatings* (Elsevier, 2010)
15. I. Khnert, M. Gedan-Smolka, M. Fischer, P. Scholz, D. Landgrebe, D. Garray, Prefinished metal polymer hybrid parts. Technol. Light. Struct. **1**(2), 89–97 (2017)

Chapter 4
Sheet Metal Tools Design

4.1 Geometric Specifications

The definition of a tool geometry is established from the part design, taking into consideration the manufacturing process to be used. Generally, the tools are defined by either cavities or cores features. Tools can be grouped into two principles. Open moulds are defined by a single tool side. The part manufacturing use only one moulding surface to shape one side of the part. The back side of the part is settle by the material thickness leading to shell type parts or by a flat plane leading to solid parts. This type of tools can either be a single part or an assembly of multiple parts, which needs to allow a feasible tool making and part release. Closed moulds use two, or more, moulding surfaces controlling both sides of the manufactured part. Again, each side of the mould can be a single part or a more complex assembly. The development of closed moulds require not only the correct definition of the moulding surface but also guiding systems for the closed mould and often clamping structures. Although these tools are more complex, they allow greater freedom for the part design due to the potential independence between the upper and lower part geometry.

Sheet metal process can be used for both tooling types. Nevertheless, it is important to consider the geometric limitations of ISF and other forming processes. Although sheet metal tools may be used for a wide variety of manufacturing processes, the geometry is mainly restricted to continuous surfaces. Thus, features like ribs or bosses are generally not feasible. Still, sheet metal tools may be used in both cavity and core moulding surfaces, although some considerations must be attended.

4.1.1 Tool Geometry Definition

As the ISF processes are better suited to the manufacturing of cavities, the definition of the geometry to be formed for negative moulds is straightforward. The side of the metal sheet facing the forming tool is used as the contact surface for moulding process. Thus, thickness variation on the sheet metal does not cause major influence on the moulded part influencing only the tool mechanical behaviour.

© The Author(s), under exclusive licence to Springer Nature Switzerland AG 2019 57
D. Afonso et al., *Incremental Forming as a Rapid
Tooling Process*, Manufacturing and Surface Engineering,
https://doi.org/10.1007/978-3-030-15360-1_4

Fig. 4.1 Geometric application of the sine law to estimate the sheet thickness after ISF and thickness influence on cavity/core features on incrementally formed sheet metal parts

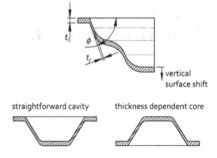

On the other hand, when developing a positive mould, as the ISF produces cavities, the reverse side of the metal sheet must be used as the contact surface of the mould. Despite controlling the tool path that forms the sheet metal, due to sheet thinning there is no precise control of the outside geometry, as illustrated in Fig. 4.1. Therefore, the definition of the mould geometry is much devious depending on the thickness. The estimation of the final sheet thickness by the sine law (Eq. 2.1) may be used to size the cavity to be incrementally formed. If the geometry requires the use of multistage forming, other estimations should be performed. When the tool development requires other forming processes for the moulding surface, different estimations should be performed. Operation where the major deformation is caused by bending may consider a constant sheet thickness without thinning phenomena. Operation where the major deformation is caused by shear may also use the sine law for final thickness estimation.

During the definition of a moulding surface, it is important to keep in mind that sheet metal forming processes, particularly ISF, have limited accuracy. Still, the right selection of the forming strategy for the parts and a right mould design may contribute to better results. Further finishing operations may be used to increase both the accuracy and surface finishing of the moulding surfaces, as presented in Chap. 3. Nevertheless, the achievable accuracy should always be a decision factor for the selection of sheet metal forming processes for the manufacture of rapid tools.

As ISF processes require a tool path programming with CAM techniques, the development of 3D models is essential for the operation. Thus, CAD software has significant relevance in the tool design and development. The CAD modelling strategy helps not only to define the entire tool design but also to geometrically estimate the final sheet thickness. While thickening operations allow to define the geometry of constant thickness sheets, it is not recommended to the ISF parts due to thinning. As an alternative, a copy and vertical movement of the moulding surface allows to geometrically use the sine law to predict thinning, and hence reaching a more accurate model with ease, as illustrated in Fig. 4.1 [1].

Analogously, the use of the right CAD tools for the definition of the models of parts manufactured by other sheet metal forming processes than ISF is also important. When a tool uses bending operations, it is crucial to use the right tools to determine the blank size, typically present in most CAD systems.

A relevant aspect about the design of sheet metal tools is related to the parting line of the part, particularly when using closed tools. Although the sheet metal tools design takes advantages of the free-form capability of ISF processes, both the tool development and the moulding operation benefits from the use of a flat parting line.

The use of a flat parting plane in closed tools or a flat support plane in open moulds benefits both the tool manufacturing and the mould operation. Although it restricts the design freedom, a flat plane simplifies the fabrication avoiding the need for trimming a joining operation leading to faster and less expensive tool development. Besides, the simpler parting line benefits accuracy, particularly on the tool closing, leading to better operation results.

The development of sheet metal moulds in rapid tooling applications is granted by ISF processes ability to form unique parts. As referred, most ISF parts are formed from a flat blank. In such a way, the use of a flat parting line leads to a straightforward tool design, benefiting both the part development time and cost, as well as minimizing the use of material. Further, the use of a flat parting line potentials the manufacture of a higher quality tool. Using a simpler part configuration allows the achievement of a more accurate shape. Besides, when using closed tools, a flat parting plane benefits the closing and sealing of the tool reducing the formation of burrs.

Notwithstanding, it is possible to use alternative shapes for the parting line. The use of pre-formed configuration parts (e.g. roll-curved sheet) allows the definition of simple curvature parting lines. However, the use of a preformed blank may lead to a more difficult forming strategy and machine setup, resulting in a slower and more expensive part development. Further, the complexity increase of the blank shape may lead to the impossibility of forming on it. The use of free-form surface for the parting line is also a possibility, taking advantage of the ISF process to define the parting surface, particularly using multistage strategies for individual features. Still, this option may require trimming and finishing operation, leading to higher complexity tools with potentially less accuracy and worst performance.

4.1.2 Tool Support

Apart from granting the right definition of the moulding surface, and, in multiple part tools, the guiding and fixing of the different parts, as well as the clamping system in closed moulds, are important to ensure the possibility of operation. As the moulding surface is defined according to the part to be formed, a similar design is found at the

Fig. 4.2 Tool support
possibilities

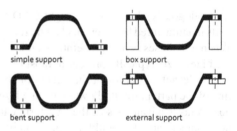

reverse side of the tools. This is often not suitable for the positioning of the tool in the moulding machine or work table. Consequently, it is important to add to the tools the possibility of an easy and stable moulding operation.

> Sheet metal tools are shell-like parts. Thus, the design of the moulding surface for a given part typically result in an inadequate outer geometry for the moulding operation. In such a way, the tool design should include the addition of supports or fixing points.

Most moulding processes require the positioning of the tools in flat tables. Therefore, sheet metal tools should be designed to allow stable support on a flat plane. Various options may be suitable to achieve this positioning, as represented in Fig. 4.2. Simple single-part tools, particularly with positive configuration may not need the use of additional support. On the other hand, negative configuration tools often need additional features to be stably placed in flat table. A simple solution is the use of a support box. Various alternatives may be used for the construction of a support box, from simple ply construction with wood derivatives to more elaborate metal work and welding fabrication. A reliable alternative is the inclusion of the support features in the tool parts by bending operations. For moulding processes where the tool is not placed on a flat table, other alternatives may be used taking advantage of external supports.

Processes which deal with several heat transfers should consider the thermal expansion for the support design. A complete clamping of the sheet metal tool may lead to a bulge effect due to thermal expansion. Thus, the addition of a flexible system either using gaps or movable fixing points may benefit the tool performance.

4.1.3   Tool Parts Guiding

An important issue about the tool design concerns the guiding of multiple parts. This may occur either for the permanent connection of two sheet metal pieces of a mould part or in the positioning of two mould parts, whether in open or closed configurations.

In the case of permanent connections between parts, the guiding could be simplified or even eliminated. These permanent links can be performed by any of the described methods in Chap. 3. Thus, depending on the used joining method, the guiding features can either be implemented directly by the part of geometric features or using positioning jigs. Permanent connections typically benefit from aligning the blank edges, therefore the formed mould cavities or cores should be placed in the same position of the blanks. Nevertheless, some mould designs require sheet metal parts trimming, removing the blank edges and making it impossible to perform simple alignments. In these cases, other alternatives must be employed. The creation of positioning drilled holes may help the lining up of parts. As an alternative, dedicated flexible jigs complementary to the mould geometry may be used to position the parts before joining.

In the case of multiple part tools, the guiding assumes a more important role since it must allow a correct mould closing in each production cycle without requiring any complicated or time-consuming assignment. The guiding features in sheet metal moulds should be implemented in a way to allow both an easy close of the mould and adjusting the closing position. Two possible approaches can be employed for the guiding, with the guiding features either backed by the support structures or placed on the sheet metal parts itself.

The guidance of multiple parts can be defined by the alignment of holes using positioning pins, by the interaction between the overall geometry of the parts to be connected or by side resting of a part against the other. The selection of the guiding method depends on both the tool geometry and the accuracy requirements. The guiding of two parts could be improved by the implementation of position tuning systems. Within these options, guidance should be granted by a fixed feature in one part and a position tuning element on the other. The backrest of the two (or more) elements grants the right positioning of the tool parts.

4.1.4 Clamping

Whenever a tool uses multiple parts, either to allow an easier demoulding in open moulds or for joining two or more parts of closed tools, a clamping system is needed. Some of these clamping needs may be granted by the used machine, requiring only a correct positioning and fixation of the tools' parts. On the other hand, most of the uses for sheet metal tooling are found in low-volume production processes, where the tool clamping must be ensured autonomously. Whichever the way, the tool must be designed to allow the permanent or disassemblable connection between parts.

Several clamping systems may be used for the connection of multiple sheet metal parts. Among other possibilities, the most common is the use of screws with nuts or riveted nuts, screw-driven c clamps, spring-loaded clamps or toggle latches. Tools like sheet metal locking pliers or welding clamps may also be used to close sheet metal moulds. Figure 4.3 presents some of the most relevant clamping options.

Fig. 4.3 Tool clamping
possibilities

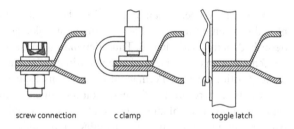

screw connection   c clamp   toggle latch

The use of screw connections between parts offers one of the most flexible solutions as it can be applied both at the outer edge of the sheet parts or at the centre, closer to the cavities or cores features. Besides, the connection elements may be orientated either along the pulling direction or in any other orientation, adapting to free-form parting lines. Further, the simple clamping system is easily applied using only high availability components. However, despite the good performance, the use of screw connections leads to a significant clamping and opening time, not being recommended for fast production cycles. Still, a right selection of the used fastenings, maximizing the use of fixed components, may reduce the operation time issues.

Other clamping systems can be used to simplify and accelerate the production rate. The use of different clamps or latches, both screw driven, toggle or spring-loaded may be used for a faster tool clamping and opening. Yet, the use of these clamping systems can lead to a longer tool manufacturing time and has a stronger influence on the tool cost. Besides, the use of more complex clamping components may require the increase of the tool structure or support. As an alternative, flexible clamping tools may be a good solution for a suitable tool clamping, leading to a short clamping time without an increase of the tool complexity.

4.2   Mechanical Behaviour

The design of a sheet metal tool is strongly dependent on the manufacturing process where it will be used, as well as the shape of the part to be manufactured. The tool geometry is defined to allow a proper moulding, regarding the guidelines related to shape, support and operation. Notwithstanding the importance of the tool geometry, the mechanical behaviour of a sheet metal tool is highly dependent on the used material and thickness. Thus, a proper analysis should be performed to foresee the tool stress and displacement under the moulding loads to state a material and thickness selection.

The use of rapid sheet metal tools finds its greater applicability in the development of moulds for low-pressure processes. Thus, the loading is often limited. Still, depending on the moulding process, the loading varies in both intensity and distribution with a typical range from 0 to 1 MPa of hydrostatic pressure.

The analysis of the mechanical behaviour of a sheet metal tool might be performed using either analytical and empirical models or computational methods. While the first methods allow a fast way to perform rough estimations, they are limited in shape complexity and only provide information about partial areas of the moulds, thus being relevant for initial material and thickness selection. The computational approach grants a more accurate and complete evaluation, being indicated for a global analysis and tool design optimization.

As sheet metal tools have a small thickness in comparison to the radii of curvature, the mechanical behaviour of the mould structure can be defined by the theory of plates and shells. Given the mould geometry, the mechanical behaviour may be described as a group of thin plates with small or large deflection and/or shells with membrane stresses, considering the possible elastic stability [2]. This rough discretization allows the use of analytical and empirical formulas under simplified loading conditions to test the mechanical response of large tool sections. A more refined discretization using computational methods permits a better definition of the loading conditions on simultaneous analysis of the complete tool.

Further, the thickness thinning must be considered to determine the mechanical response of the tool under the moulding loads. For this, the approximation given by the sine law provides a fair assessment for the stress and deflection calculation. As a consequence, higher stress zones are often placed in the thinner areas of the tools compromising its potential performance. Yet, some considerations must be taken into account for the analysis of the mechanical behaviour. On one hand, the stress analysis should consider the material strain hardening to compare against the determined stress [3]. On the other hand, if the stress overtakes the yield limit or the deformation exceeds the acceptable values, the tool may be reinforced or further supported to improve its performance.

4.2.1 Analytical Analysis of the Sheet Metal Behaviour

As referred, the mechanical response of a sheet metal tool under a moulding load may be determined using plates and shells theory. However, due to the great variety and high geometric complexity, the analytical analysis is not simple. Still, some analytical and empirical formulas may be used to forecast the mechanical behaviour of some areas of the tools. This approach does not allow to understand the full performance of the tools, but can be used for an initial material selection and estimation of the proper thickness.

The use of formulas to determine the mechanical response of the tools should consider areas of the sheet metal parts with approximated shape of simple case studies. Flat areas of the tool may be analysed as cases of plates of constant thickness. Depending on the wall slope, the thickness may be obtained from the initial blank or from a constant application of the sine law. Curved or double curved areas may be considered as shells. In these cases, the thickness varies due to the slope-dependent thinning. Thus, an average value may be estimated or the worst-case scenario considering the maximum thinning may be used for the mechanical evaluation.

Table 4.1 presents some simple case studies with the potential to be used in the mechanical analysis of the mechanical response of sheet metal tools sections under the moulding loads. The formulas consider a uniform loading over the entire plate or shells p in N/mm^2, dimensions in mm, material Young modulus in MPa. All plate cases are considered with the plates clamped in all edges and the shells fixed in the bottom. The calculus results are given in mm for displacement (δ) at the centre and MPa for stress (σ_c) at the centre and (σ_e) at the edge [4, 5].

> The use of analytical and empirical equations allows a fast initial forecast about the behaviour of a mould section under simple loading conditions. Although this approach is not sufficient to validate the full mould design, it supports the initial material and sheet thickness selection.
>
> Depending on the tool shape, the analytical approach should be applied to the largest flat areas of the tools, more likely to suffer larger stress and displacement. Different equations sets are suitable to approximate the behaviour of particular tool shapes and more than one case may be used for one tool analysis.

The first three lines from Table 4.1 represent examples of shapes that can be used to simulate flat areas on the top or bottom of the tools. The plates are considered clamped all around to represent the material continuation. Regarding the tool shape, a rectangular, circular or elliptical panel may be used to reach the better fitting. The maximum displacement is determined at the panel centre and the stress value is determined both at the centre and next to the clamped edges. The three equation sets are determined from plate theory with small displacements. Thus, results are only considered accurate if the maximum displacement is inferior to the thickness, either the blank or the thinned sheet. Larger displacements may only be seen as an estimation.

The plate under compression represents a suitable case to analyse the side panel of a mould, which, additional to the hydrostatic pressure, supports the load from the upper part of the tool. These loading schemes add membrane forces to the panel leading to supplementary compression stress and strain. However, this loading conditions may lead to a buckling effect of the flat panel. In such a way, this set of equations presents the determination of a critical stress value, from which it is likely to occur bucking.

The last two lines of Table 4.1 presents equations to determine the stress and displacement on shells with spherical and cylindrical shape with tangential edge support under uniform loading. The cylindrical shell is considered with tapped ends to represent the effect of material continuation. Due to the slope-angle-dependent thinning, it is common to have variable thickness shells. As the calculus formulas are stated for constant thickness, two approaches are possible. Considering the thinner section as a reference ensures a safe approach, avoiding the calculation of stress

Table 4.1 Formulas for plates and shells under stress [5]

Case	Deflection	Stress
rectangular plate	$\delta_{max} = \dfrac{-\alpha p b^4}{E t^3}$ $\begin{array}{l\|lllllll} a/b & 1.0 & 1.2 & 1.4 & 1.6 & 1.8 & 2.0 & \infty \\ \hline \alpha & 0.0138 & 0.0188 & 0.0226 & 0.0251 & 0.0267 & 0.0277 & 0.0284 \\ \beta_1 & 0.3078 & 0.3834 & 0.4356 & 0.4680 & 0.4872 & 0.4974 & 0.500 \\ \beta_2 & 0.1386 & 0.1794 & 0.2094 & 0.2286 & 0.2406 & 0.2472 & 0.2500 \end{array}$	$\sigma_{e\,\text{center of long edge}} = \dfrac{-\beta_1 p b^2}{t^2}\ (\sigma_{max})$ $\sigma_c = \dfrac{\beta_2 p b^2}{t^2}$
circular plate	$\delta_{max} = \dfrac{-3 p r^4 (1-v^2)}{16 E t^3}$	$\sigma_e = \dfrac{-3 p r^2}{4 t^2}\ (\sigma_{max})$ $\sigma_c = \dfrac{3 p r^2 (1+v)}{8 t^2}$
elliptical plate	$\delta_{max} = \dfrac{-3 p b^4 (1-v^2)}{2(3+2\alpha^2+3\alpha^4) E t^3}$ $\alpha = \dfrac{b}{a}$	$\sigma_{ea} = \dfrac{-6 p b^2 \alpha^2}{(3+2\alpha^2+3\alpha^4) t^2}$ $\sigma_{eb} = \dfrac{-6 p b^2}{(3+2\alpha^2+3\alpha^4) t^2}\ (\sigma_{max})$ $\sigma_{ca} = \dfrac{3 p b^2 (\alpha^2+v)}{(3+2\alpha^2+3\alpha^4) t^2}$ $\sigma_{cb} = \dfrac{3 p b^2 (1+v\alpha^2)}{(3+2\alpha^2+3\alpha^4) t^2}$
panel under compression	$\delta = \dfrac{pa}{Ebt} + \delta_{\text{buckling}} + \delta_{\text{Out-of-plane Actions}}$	$\sigma = \dfrac{p}{tb} + \sigma_{\text{Out-of-plane Actions}}$ critical stress: $\sigma' = K \dfrac{E}{1-v^2} \left(\dfrac{t}{b}\right)^2$ $\begin{array}{l\|llllllllll} a/b & 0.4 & 0.6 & 0.8 & 1.0 & 1.2 & 1.4 & 1.6 & 1.8 & 2.1 & \infty \\ \hline K & 7.76 & 5.80 & 6.00 & 6.32 & 5.80 & 5.76 & 6.00 & 5.80 & 5.76 & 5.73 \end{array}$
spherical shell	$\delta_{max} = \dfrac{p r^2 (1-v)(1-cos\theta)}{2 E t}$	$\sigma_1 = \dfrac{p r}{2 t}$ $\sigma_2 = \sigma_1$
cylindrical shell	$\delta_{max} = \dfrac{p r l (0.5-v)}{E t}$	$\sigma_1 = \dfrac{p r}{t}\ (\sigma_{max})$ $\sigma_2 = \dfrac{p r}{2 t}$

values lower than the real. Although, this approximation may lead to an oversized sheet material or thickness selection. As opposite, the use of an average value for the thickness result on a better approximation at the cost of underestimating the stress and displacement. In these shell structures, the stress and displacement are largely uniform at the points away from the edges. The spherical shape features a uniform stress distribution. On the cylindrical shell, the maximum stress occurs along the circumferential axis double than the longitudinal axis.

4.2.2 *Numerical Calculation*

When the complexity of geometries or materials employed impair the use of analytical tools, the use of computational methods provide an ideal framework for tooling design requiring a minimum of time and spending no material. Among the numerical tools used in structural calculation, the finite element method (FEM)[6] has become the most widespread one, being accessible through many commercial packages like Abaqus, Ansys, LS-Dyna, etc.

In short, for a physical problem governed by a set of differential equation and boundary conditions, the FEM divides the geometric domain into smaller domains (the so-called finite elements and its nodes) and equilibrium is achieved for each finite element node. The differential equations are then transformed into a system of algebraic equations that can be integrated over time based on two different integration schemes: explicit, when inertial effects are particularly important (e.g. impact tests) or implicit (quasi-static analysis). Both solution procedures are commonly available in commercial FEM codes.

The explicit integration is a dynamic approach, which gives the equation system solution without a request for an iterative procedure. An algorithm scheme uses a diagonalized mass matrix and the final force balance is not checked. The differences between the internal and external forces are used to calculate the nodal acceleration, velocity and displacement. The nodal positions at the end of the step are extrapolated using the initial nodal position and acceleration field. This method is conditionally stable, which means the solution converges as long as the time increment size is smaller than a critical value. The most important advantage of the dynamic explicit scheme besides handling mass effects is the faster CPU time. However, it is more prone to accuracy issues.

In the implicit approach, the static equilibrium of the algebraic equations is satisfied at the unknown final configuration of a time increment. This method enables a full static solution of the deformation problem with convergence control. Theoretically, the increment sizes can be very large, but they can be limited due to the contact conditions. The finite element equation solution in implicit integration scheme involves an iterative procedure to achieve the convergence criterion at each increment. This iterative procedure is based on Newton–Raphson method. The major advantage of the implicit method is its unconditional stability: It can provide a correct solution independent of the time step/increment, but CPU times can be much higher compared to explicit approaches.

In this sense, FEM simulations have been intensively used in tooling design (static analysis) and also in simulation of forming processes (dynamic or static) in order to better predict the structural behaviour during any component forming and its final geometry.

An accurate prediction of the forming forces during manufacturing process is also of critical nature as it contributes to a safe, long-lasting tooling and hardware use. Particularly for SPIF, the forming forces prediction is particularly important in the case of using adapted machinery not designed primarily for the SPIF process.

4.2.3 Tool Reinforcement and Back Support

The use of thick sheets or high-strength materials has the potential to develop standalone sheet metal tools. Notwithstanding a lot of applications loading requirements are compatible with standalone sheet metal moulds, other cases require additional strength. These cases occur either when dealing with high moulding loads or when the mould forming operation limits the use of stiffer and stronger materials or thicker thicknesses.

Still, the impossibility of developing strong enough tools does not disrupt the use of rapid sheet metal tools. The use of reinforcements and local or global support structures may increase the application field of sheet metal moulds.

On the first stage, the tool support structure, as presented in Sect. 4.1.2, may be sized accordingly to the moulding loads. The position of support boxes, bent supports or external supports may be placed to minimize the width span of the unsupported area. Although this effort may increase the geometric complexity of those support structures, it reduces the unsupported area under pressure, reducing not only the width span but also the total load. Besides, the additional downside is the decreased possibility of reusing support boxes or external structures for multiple tools. A smart option to improve the boundary support of the sheet metal parts, particularly when using single-part moulds manufactured in SPIF, is the reuse of the backing plate as a support structure. Tools developed from the permanent joining of multiple parts may also explore this possibility at the cost of additional trimming, forming and joining operations of the backing plate parts.

As an alternative, sheet metal parts may be reinforced in specific points or use a general reinforcement or back support.

The development of a grid of supports under the sheet metal due to severe decrease of the suspended area leads to a strong increase of the tool stiffness and strength. The reinforcement grid may be applied under the overall area of the tool or in particular extents. Different materials and building techniques may be used for the development of these grids, with major relevance to the use of metal sheets or wood derivate boards. Though these options may fulfil the loading requirements of the tools, they may assume a significant weight in the mould development and manufacturing [7]. In such a way, these options are restricted to the development of moulds with simple shapes in rapid tooling operations. As an alternative for tool reinforcement, the use of gussets assumes a similar goal while being a simpler solution, typically with easier and faster implementation.

A different possibility for a global back support is the use of composite materials as reinforcements for the sheet metal structure. Hand moulding techniques such as hand lay-up or spray-up, mainly using chopped or recycled fibres allow the strengthening of the sheet metal parts, without the need for additional tools [8]. These strengthening options are mainly interesting for the development of large tools, and may be allied to the use of gussets of support grids. Nonetheless, as this reinforcement method has a continuous bond to the sheet metal, its use may be limited in heat-dependent processes because of different thermal expansions.

An alternative to the use of reinforcements is the addition of complete back support. When using peripheral support box structures for the tools handling, it is possible to fill up the tool core to increase its strength and stiffness. The use of resins, either dense or foams, may be suitable for these filling. However, this option leads to a permanent, typically non-recyclable, use of the material, thus not very interesting mainly on medium to large volumes. As an alternative, the use of a porous mixture also fulfils the filling and support purpose, with the possibility of being reused [7]. Some alternatives for filling materials are sand, clay pellets or metallic spheres, pellets or coarse powder. The bottom closing of the support box seals the filling mixture inside the tool core and allows its use, being compatible with temperature variations and allowing vacuum operations in a drilled tool. An interesting alternative for the reinforcement of the sheet metal tools uses a pressurized air or fluid, which may also benefit the thermal behaviour [9].

Apart from the presented strategies, new approaches using hybrid manufacturing processes have great potential for the development of locally reinforced sheet metal parts.

A promising technique is the use of laser additive manufacturing over the sheet metal parts to create local reinforcements. The flexible addition of material allows the development of lightweight smart tools, with tailored design reinforcement to reach the most adequate mechanical behaviour [10].

The use of tailored blanks, either from overlapping sheet metal patches or by welding different thicknesses sheets in a design mat allows an uneven thickness distribution of the final part [11, 12]. This over thicknesses allow reinforcing particular areas of the tools, making them better suited for the moulding operation.

4.3   Thermal Behaviour

The thermal behaviour of the ISF tooling parts has a strong influence on the moulding processes, mainly in heat-related technologies. Due to the reduced mass, the ISF parts have low thermal inertia which benefits the heating and cooling cycles, contributing to a faster moulding process. However, the thinning during the ISF result on an uneven thickness distribution of the sheet metal parts. These thickness differences lead to potential dissimilarities in the heating ratio of the tool areas, which may harm the moulding process.

As in the mechanical analysis, the thermal behaviour can be performed using the thickness prediction by the sine law. Sheet metal parts that are obtained from other forming processes besides ISF may be considered to have constant thickness or thickness variation determined by other methods. Tools that are finished with coatings may issue the worst thermal behaviour as the filling material could cause local over thicknesses that harm the temperature distribution.

The heating and cooling of the sheet metal tools can be performed essentially using two methods. The moulds can be heated through forced convection in an oven or can be directly heated through conduction using resistors or a heating fluid circuit.

Similarly, the cooling can also be performed through convention at room temperature or under forced vending or water showers or by conduction with a cooling fluid circuit.

The thickness differences along the tool are potentially more unfavourable in convention heat/cool as all the tool surface is typically under the same heating conditions. Still, heat sinks or thermal insulation can be used to speed up or reduce the heat ratio. When the tools' temperature is controlled using resistors or heating fluid, their distribution can cancel the thickness variation heat rate differences along the sheet.

The calculus of the heating time and temperature distribution has an important role in the tool design. When dealing with conduction heated/cooled tools, the thermal analysis contributes to the definition of the heating grid. When dealing with convention heated/cooled tools, the thermal analysis has a major applicability in validating the temperature differences and sizing thermal insulation or heat sinks. In both scenarios, the heat/cool ratio calculus has an important role in forecasting the moulding cycle time.

Apart from forecasting the heat/cool cycle time, the thermal analysis has also a strong influence on the mechanical aspects. The temperature variation causes the material to change shape due to thermal expansion. Depending on the tool geometry and constraints, this expansion may lead to relevant reshaping or cause significant inner stresses. The thermal expansion and thermal stress must be considered to validate the proper operation of the tools. Besides, processes that deal with moderate to high temperatures may have a significant role in the tool itself, as the material faces a heat treatment during the first moulding cycle. In these scenarios, the heat treatment should also be considered and the material should be subjected to heat treatment before first use.

4.3.1 Analytical Analysis

The heat up cycle analysis thought analytical methods allows a fast estimation on the reachable temperature for each mould panel for a given heating time, or estimate the heating time to reach the desired temperature. The analytical approach is mainly useful for the temperature evaluation of tools heated up by convection, which occurs in the processes where the sheet metal parts are heated inside an oven. In this case, the heat transferred per unit time is given by Eq. 4.1, where h is the forced convection coefficient, A the surface area and dT the temperature difference between the surface and the bulk fluid.

$$q = h.A.dT. \tag{4.1}$$

By integrating the convective heat gain rate over time, one can determine the surface temperature over time or the heating time to achieve a given temperature by Eq. 4.2, where Δt is the mould heating time, $\rho.c_p$ is the volumetric heat capacity, T is the mould wall intended temperature, T_0 is the oven temperature and T_i is the mould initial temperature. V represents the mould material volume and A the surface area,

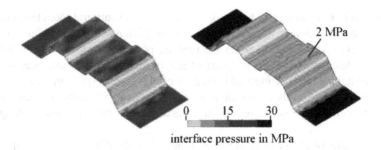

interface pressure in MPa

Fig. 4.4 Thermomechanical analysis of a sheet metal formed product, [17]

where $V/A = t$, the mould thickness. Though wall thickness temperature gradient can be neglected since resistance to heat conduction for a thin-walled mould is much smaller than convective heat flow resistance. The plastic material heat is slower due to the higher heat flow resistance and the energy used for the change of state [13, 14].

$$T = T_i + (T_0 - T_i) \times \left(1 - e^{-\frac{h.A}{\rho.c_p.V}.\Delta t}\right) = T_i + (T_0 - T_i) \times \left(1 - e^{-\frac{h}{\rho.c_p.t}.\Delta t}\right)$$

$$\Leftrightarrow \Delta t = -\frac{\rho.c_p.V}{h.A}.\ln\left(1 - \frac{T - T_i}{T_0 - T_i}\right) = -\frac{\rho.c_p.t}{h}.\ln\left(1 - \frac{T - T_i}{T_0 - T_i}\right)$$

$$(4.2)$$

In addition to the reaching temperature and the cycle calculation, analytical models can also be used to test the expected thermal expansion behaviour.

4.3.2 Numerical Simulation

Despite being first applied to structural analysis, one of the great advantages of finite element approximations is the ability to adapt to any kind of physical phenomena, as long as its governing differential equations and boundary conditions are known a priori [15]. In this sense, thermal problems can be analysed in full analogy regarding structural ones, being the primary variable the temperature at elements nodes (instead of displacements), loads being heat fluxes (instead of forces and moments), stiffness being replaced by conductivity and so on. Moreover, it is possible to couple temperatures and displacements in the same analysis framework, achieving thermal and static equilibrium simultaneously [16]. This technique is extremely important to evaluate thermomechanical effects whenever cyclic thermal cycles can reduce tooling lifetime or when the manufacturing process required heating and/or cooling to achieve ideal formability conditions like in hot forming of high-strength alloys, Fig. 4.4.

References

1. D. Afonso, R. Alves de Sousa, R. Torcato, Defining design guidelines for single point incremental forming—rules to a good design of container and tunnel like parts manufactured by incremental forming, in *Challenges for Technology Innovation, An Agenda for the Future, 2017*, pp. 195–200
2. S. Timoshenko, S. Woinowsky-Krieger, *Theofy of Plates and Shells* (McGraw-Hill, New York, 1959)
3. W. Emmens, A. Boogaard, *Formability in incremental sheet forming and cyclic stretch-bending* (Steel Research International, Wiley, Weinheim, 2011)
4. M. Qatu, E. Asadi, W. Wang, Review of recent literature on static analyses of composite shells: 2000–2010. Open J. Compos. Mater. **2**(3), 61–86 (2012)
5. W. Young, R. Budynas, *Roark's Formulas for Stress and Strain* (McGraw-Hill, Singapore, 2002)
6. O.C. Zienkiewicz, *The Finite Element Method in Engineering Science* (McGraw-Hill, London, 1971)
7. R. Appermont, B. Mieghem, A. Bael, J. Bens, J. Ivens, H. Vanhove, A. Behera, J. Duflou, Sheet-metal based molds for low-pressure processing of thermoplastics, in *Proceedings of the 5th Bi-Annual PMI Conference*, vol. 5 (2012), pp. 383–388
8. I. McColl, J. Morley, Damage tolerant fibre reinforced sheet metal composites. Philos. Trans. R. Soc. Lond. Series A Math. Phys. Sci. **287**(1338), 17–43 (1977)
9. J. Bens, B. Mieghem, R. Appermont, H. Vanhove, A. Bael, J. Duflou, J. Ivens, Development of material-and energy-efficient metal sheet based tools for composite manufacturing, in *Proceedings of the ECCM15—15TH European Conference on Composite Materials* (2012)
10. M. Bambach, A. Sviridov, A. Weisheit, J. Schleifenbaum, Case studies on local reinforcement of sheet metal components by Laser additive manufacturing. Metals **7**(4), 113 (2017)
11. M. Merklein, M. Johannes, M. Lechner, A. Kuppert, A review on tailored blanks—production, applications and evaluation. J. Mater. Process. Technol. **214**(2), 151–164 (2014)
12. M. Silva, M. Skjoedt, P. Vilaa, N. Bay, P. Martins, Single point incremental forming of tailored blanks produced by friction stir welding. J. Mater. Process. Technol. **209**(2), 811–820 (2009)
13. H. Belofsky, *Plastics: Product Design and Process Engineering* (Hanser, New York, 1995)
14. S. Banerjee, W. Yan, D. Bhattacharyya, Modeling of heat transfer in rotational molding. Poly. Eng. Sci. **48**(11), 2188–2197 (2008)
15. J. Fish, T. Belytschko, *A First Course in Finite Elements* (Wiley, 2007)
16. A.E. Tekkaya, H. Karbasian, W. Homberg et al., Prod. Eng. Res. Devel. **1**, 85 (2007). https://doi.org/10.1007/s11740-007-0025-9
17. L. Penter, S. Ihlenfeldt, N. Pierschel, IOP Conf. Ser. Mater. Sci. Eng. (2018). https://doi.org/10.1088/1757-899X/418/1/012012

Chapter 5
ISF Rapid Tooling Applications

5.1 Geometry Definition and Tooling Predesign

The authors have developed some fundamental research on SPIF-based RT techniques involving the study of moulds for different manufacturing processes. For each hypothesis, the research consists of the mechanical and thermal design of the mould, the forming and assembly of the mould itself and the test of the moulding operation. The mould manufacturing process and operation performance was analysed, evaluated and, when possible, compared against conventional tooling.

A reference geometry was designed to support the RT development for different technologies. As the research work aimed for a proof of concept, the reference geometry was designed to ensure a feasible forming operation by a single-stage SPIF tool path strategy. A reference geometry derives from a drafted volume with a maximum wall angle of 70°. To avoid a full symmetry, one sidewall is sloped at a smaller angle. The part is sized to fit in a 200×200 mm forming window for operational convenience.

The reference geometry is designed with flat areas in order to cut test specimens to use in tensile tests. Since the rapid tooling research seeks the development of thermoplastics and thermoset materials, type IV specimens [1] with overall dimensions of 115 mm by 19 mm are chosen to be used in a direct comparison between materials. This allows not only to analyse the moulded material but adds the potential to compare it against the use of conventional tooling. For the part design, two flat areas are chosen with at least 120 mm by 25 mm to cut the specimen for tensile tests, as presented in Fig. 5.1.

Since the research aimed to develop hard tooling techniques, the material properties used in the mould-making have a strong influence on its performance. Due to its high formability and to the great availability, AA1050 H111 aluminium sheet is selected for the tool development. The alloy is a popular grade of aluminium for general sheet metal work where moderate strength is required. Sheets can be found between 0.2 and 6.0 mm. The alloy temper is annealed and slightly strain-hardened, given a minimum yield strength of 85 MPa and an ultimate strength between 105 and 145 MPa. The sheet has a density of $2710 \, \text{kg/m}^3$, a Young Modulus of 71 GPa and a Poisson ratio of 0.33.

© The Author(s), under exclusive licence to Springer Nature Switzerland AG 2019   73
D. Afonso et al., *Incremental Forming as a Rapid
Tooling Process*, Manufacturing and Surface Engineering,
https://doi.org/10.1007/978-3-030-15360-1_5

Fig. 5.1 Reference
geometry for the rapid
tooling development

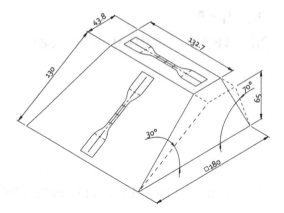

To analyse the influence of the forming operation on the material, tensile tests were performed to measure the true yield after strain. Plate type 1 specimens [2] were cut from an incrementally formed part shaped from the reference geometry. Three different sets of specimens were cut: from an unformed sheet, from the 30° slope wall and from the 70° slope wall. The analysis of the mechanical properties of the material at each strain hardening state is a key element for the mechanical analysis of the tools an appropriate mould thickness sizing. The cut of the specimens not only allowed to characterize the material but also confirmed the final thickness at the differently formed angle of the part and compare it against the sine-law estimation, presented in Table 5.1.

For the fulfilment of mechanical characterization tests, three parts based on the reference geometry were formed out of AA1050 H111 2 mm sheet. The edge between walls was filleted with a 20 mm radius to allow the forming operation to run at 3000 mm/min without machine vibration issues. The forming operation used a 12 mm spherical punch in a single-stage helical tool path strategy with a 0.5 mm vertical step down. The forming operation was completed without any sheet damage apart from a slight skinning. Three sets of specimens were cut from the three different parts, registering the same thickness in all parts. The specimens cut was performed on a five-axis CNC machine using a 6 mm milling tool operating perpendicular to the sheet. The specimens were finished by deburring, sanding and polishing the trim cuts before performing the tensile tests.

Table 5.1 Formed part wall
thickness

Part area	Estimated thickness
Unformed top	2.00 mm
Smaller slope wall	1.73 mm (−13.5%)
Higher slope wall	0.68 mm (−66.0%)

Fig. 5.2 Mechanical
response of a 180 × 130 mm
rectangular 1.73 mm
thickness and a
180 × 69 mm rectangular
0.68 mm thickness
aluminium alloy sheet under
uniform pressure

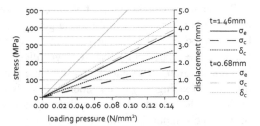

From a geometric point of view, the cut of the specimens and thickness measurement validates the use of the sine law for the prediction of the thinning effect in the AA1050 sheet. The sine law estimates a thickness reduction of 13.5% on the 30° slope wall and 66% on the 70° slope wall and the measure registers a respective thickness reduction of 13.2 and 68.4%. This small difference supports the use of the sine law for the thickness sizing of tools for different technologies.

The tensile tests were performed in a Shimadzu AG-IS 10kN universal testing machine at 5 mm/min. The test on the thicker specimens takes just over 30 s to break, the specimens from the smaller slope walls take close to 40 s to break and the specimens from the higher slope wall take only just over 10 s to break. The specimens cut from the unformed areas fail by neck down break and the ones cut from the formed areas tend to break at 45°. The specimens cut from the higher slope break almost without additional permanent stain. In the remaining specimens, a material elongation is visible in the tested specimens.

The yield value on the unformed specimens is 103 MPa. In what considers the specimens from the smaller slope, no significant variation is found. On the other hand, the yield occurs at 130 MPa on the specimens cut from the higher slope. In addition, the thinner specimens' material stiffness appears to be slightly higher and breaks without any plastic deformation since the forming operation already induced the maximum possible strain hardening. The evaluation of the mechanical properties of the material, particularly the perception of an increase in the yield value from $\sigma_y = 103$ MPa to $\sigma_y = 130$ MPa, supports the mechanical design of hard rapid tooling. Although it is expected that thinner walls are under higher stress values, understanding yield behaviour after strain hardening is crucial for a thickness sizing of a rapid tooling mould.

Considering the reference geometry, two possible sides might face considerable stress values. The smaller slope features the largest flat area, close to 180 × 130 mm with an expected thickness of 1.73 mm. The opposite side has a smaller area with about 180 × 69 mm but a reduced thickness of 0.68 mm.

For the major part of the manufacturing processes, the tool walls are under uniform hydrostatic pressure. Thus, the mechanical behaviour of the flat, near rectangular walls may be described by the equations of the first line of Table 4.1. The graphic of Fig. 5.2 presents the stress and displacement on the approximated flat areas of the sheet metal part under hydrostatic pressure.

The analysis of rectangular flat plates' behaviour estimates a maximum load of only 0.05 N/mm² (0.5 atm) on the larger area and 0.025 N/mm² (0.25 atm) on the higher slope wall due to the stress at the clamped panel edges, although, as all manufactured parts must feature edge fillets, a stress concentration at the edges is reduced, along with and decrease of the flat area. Thus, a higher value for the maximum loading may be considered. In such a way, considering a reduced area by excluding the cylindrical fillet, and giving major attention to the stress at the centre of the panels, a plausible maximum loading of 0.1 N/mm² (1.0 atm) is expected.

The use of 2 mm thickness 1050 aluminium alloy allows developing stand-alone tools for moulding processes with pressure up 1 atm for parts based on the reference geometry. The addition of supports increases the application field of these tools. This low-strength alloy provides a basic benchmark for the rapid tooling development, as other materials lead to better performances.

Apart from the mechanical behaviour of the tool, the thermal inertia is an important issue on sheet metal tools, particularly in heat-related processes. In those processes, since the material heating occurs mainly by heat conduction from the mould surface, the uniform heating of the mould benefits the moulded part. However, as referenced, ISF moulds have uneven thickness distribution due to thinning. Thus, an evaluation of the mould heating time and temperature distribution was performed to validate the ISF mound concept.

According to Eq. 4.2, it is possible to determine the mould heating time by convection. The preliminary analysis was performed aiming to a desired tool temperature of 200 °C, typical from several manufacturing processes. Considering the aluminium properties $\rho = 2705 \, kg/m^3$, $c_p = 1386 \, J/(kg.K)$ and $h = 20 \, W/(m^2 K)$ for aluminium on air, a mould with a wall thickness of 2 mm (t = 0,002), takes around 570 s to heat from 20 to 200 °C in a 250 °C oven. The graphic in Fig. 5.3 represents the heating time up to 200 °C in an oven from 200 to 300 °C of different thickness aluminium plates. The thinner walls on the mould with 0.68 mm thickness take only around 200 s of heating time on the same conditions. However, the heat conduction along the mould walls contributes to more balanced heating. In an ideal setting, the mould heating time could be estimated by Eq. 4.2 considering an average $V/A = t_{average} = 1.46$ mm, leading to an average heating time of 420 s in the same conditions.

In addition, Eq. 4.2 can also be used to determine the expected mould temperature after a certain amount of heating time. This evaluation can be useful to predict the mould temperature in the thinner areas at the moment when thicker walls reach 200 °C. When heated in a 250 °C oven for 570 s needed to reach 200 °C in the 2 mm walls, the thinner walls of the mould may reach almost 250 °C. The average mould temperature rounds 220 °C. The graphics in Fig. 5.4 represents the mould temperature over time in a 250 °C oven.

Fig. 5.3 Aluminium sheet metal mould heating time up to 200 °C in a forced convention oven

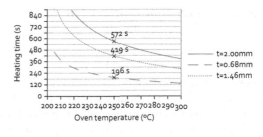

Fig. 5.4 Aluminium sheet metal mould temperature in a 250 °C forced convention oven

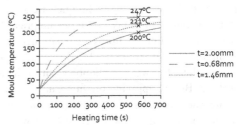

The thickness differences along the tool, due to thinning, harm the temperature distribution along the sheet metal parts. Convention heating with oven temperatures close to the desired mould temperature minimizes the thermal gradients along the tool, at the cost of longer heating times. Still, the heat conduction along the material thickness may have a slight benefit in the temperature differences, which also decrease when the tool is heated in contact with the moulding material.

5.2 Composite Materials Processing

The use of the ISF as a rapid tooling process as a particular interest in the development of tools for composite materials. These are often used in low-volume production where the tool cost is highly significant. Besides, several composite materials manufacturing processes have design guidelines compatible with ISF parts, becoming compatible with the sheet metal part design limitations.

As analysed, the use of the 2-mm low-strength aluminium sheet limits the loading capacity of the developed tools. Still, two different manufacturing processes are tested: exploring both manual operations where the loading pressure is almost negligible and close mould operation with relevant forces. Nevertheless, the part design includes not only the addition of fillets between faces but also reinforcement ribs which increases the moment of inertia of both the mould and the moulded parts.

5.2.1 Open Tool Contact Moulding

Open mould thermoset composite processes produce components with a good strength-to-weight ratio, with a fair design flexibility and a low to moderate tooling cost. The basic techniques generally use room temperature low-pressure cure of low-viscosity resins, shaping reinforcement fibres to the parts' geometry. A large variety of open mould processes can be used for the fabrication of composite parts, using both dry fabrics and wet resins or preimpregnated fabrics denoted prepregs. The most common moulding techniques are spray-up and hand-lamination, also known as hand layup or wet layup. Figure 5.5 presents the basic hand layup process. In the hand layup technique, fibres are positioned on or into the mould and wetted by liquid resin. Brushes are often used to distribute the resin evenly onto the fibres and rollers are employed to work air bubbles out of the reinforcement and to ensure complete wet out. Before applying the fibres and resin, the moulding surface can be prepared using release agents or release films, and covered in gel coat. The process is highly skilled dependent and only suited to low production rates due to slow cure times of room temperature resins. The contact moulding techniques can be improved by applying pressure to improve consolidation of the laminate by extracting excess resin and included air. This is mainly achieved using vacuum bagging, covering and sealing the hand-laid laminate inside a vacuum film over a peel ply, a perforated release film and a breather/bleeder fabric [3, 4].

> Processes like hand layup and spray-up are widely used in open tool contact moulding. The use of an open tool helps the material positioning using low-tech methods which are interesting for prototyping or low-volume production. Resins are impregnated by hand into fibres and cure, taking on the mould shape. The tool geometry allows the control of one side of the moulded part.

Different approaches can be used for positioning the reinforcement material, allowing a proper suiting to the mould surface. On the one hand, tapering pieces of fabric allow great structural tailoring capabilities, damage tolerance and potential for creating significant weight savings. However, stress concentrations at the

Fig. 5.5 Hand layup operation principle

drop-offs of the fabric pieces may lead to delamination in resin-rich areas referred to as resin pockets. The stacking of the reinforcement plies, both in number of layers and positioning, has strong influence on the part properties [5]. On the other hand, the fabric can be sheared, allowing a better fit to the mould surface. Different techniques like one-handed guiding or two-handed guiding, manual folding, hoop shearing, double-tension shearing, tension-secured shearing and mould interaction shearing both before and during the fabric positioning allow a better adjust on the mould surface [6].

The materials used as reinforcements are typically aramid, carbon or especially glass fibres. This is available in chopped strand mat and in woven, knitted, stitched or bonded fabrics. The most used thermosets in contact moulding are polyester, vinylester and epoxy resins. The moulding techniques to use are selected depending on the intended part quality and on the selected materials. The curing procedure is also selected depending on the thermoset resin, either at ambient temperature or autoclave curing [3].

The moulds used for contact moulding are found in both a core and cavity configuration, where the moulding process aims to control the accuracy and surface finishing of the inner or outer face. The tool should have adequate rigidity to maintain dimensional tolerance and a surface finish to reflect that required of the component to be produced. The moulds used for contact moulding are commonly manufactured in-house, using fibreglass-reinforced skin, and supported by a framework, commonly welded steel structure or wood section box. These are typically manufactured from models of the final part built in wood, plaster, foam or other materials. The master is completed, polished and waxed and the mould is built upon it with fabrication technique similar to the parts manufacturing [3].

Depending on the part complexity and size, the mould can be a single part or a complex assembly. However, regardless of the mould complexity, the moulding techniques produce continues surfaces. Thus, the contact moulding techniques must be followed by trimming, drilling or other finishing operations. Although the single-side moulds only allow to control the dimensional tolerance and surface finishing of one-part side, it is possible to smooth out and gel coat paint the non-tool side.

Despite the possibility of developing low to moderate cost tools, the manufacturing process is typically long-standing and commonly generates non-recyclable waste. As the moulding pressure is low, it is possible to replace conventional moulds by sheet metal moulds, manufactured by ISF processes. The study hypothesis tests the use of SPIF for the development of a cavity mould to be used in a fibreglass hand layup process. The mould is developed in a single part, formed from a single aluminium flat blank. Since sheet metal rigidity is sufficient to support the moulding pressure, the mould dispenses the need for a framework, being mounted only on a simple support to ensure stability.

Since the moulding force is only hand applied and yet low, the load support capability is not an issue. The 2-mm-thick sheet both allows a feasible forming processes using a single-stage operation and provides great mechanical performance for the moulding operation.

The manufacturing process of the hand layup mould started with the cut of a backing plate using a water jet. The forming operation was performed using a 12 mm spherical punch, with forming step of 0.5 mm in a single-stage helical tool path strategy. Because of the short moves due to the small indents, the forming operation was performed at a 2500 mm feed rate, avoiding vibration-related issues. The process took just over an hour and a total energy consumption of 6.5 kW.h including close to 40 min of effective forming operation. For the tool support, a simple MDF box was built out from four equal pieces, assembled by side screwing. Total fabrication time for the mould was 1 h 38 min being 67% for the SPIF process. The total energy consumption was 9.13 kW.h, although a large proportion was derived from the water jet cutting. The total material cost was just 7.00€, being 2.50€ for the backing plates, 3.50€ for the aluminium sheet and 1.00€ for the MDF parts and fasteners.

After manufacture, SPIF sheet mould was measured by contact using a touch-trigger probe on a 3-axis coordinate measuring machine (CMM). Maximum deviation is +6.4 mm on the sidewalls of the mould, with an average deviation of +1.5 mm. The measurement analysis shows an accuracy improvement at the indent-reinforced walls. When analysing the reinforced walls and part bottom only, maximum deviation is −3.0 mm at the part top, with an average deviation of +0.9 mm.

The mould operation test was made in a hand layup process using 400 g/m^2 fibreglass chopped strand mat impregnated with epoxy resin. The part was laminated in a three plies composite, using mainly tapering pieces techniques with two-handed guiding and manual folding.

The mould was spray covered with silicone-based demoulding agent. No gel coat was used for the part manufacture. The fibreglass mat was tailored so that each layer can cover different areas of the part without cuts. The first and the third plies used a larger mat piece extended from the top of the smaller slope wall, by the parts bottom to the top of the opposite wall and two smaller pieces on the remaining sidewalls. Small pieces of mat were also placed in the radii between sidewalls. The second ply had a perpendicular mat distribution.

All pieces of fibreglass were soaped in resin before being assembled in the part lamination. The pieces were positioned one at a time and pressed against the mould using a barrel-type roller and hand pressure. The mould rigidity was plenty sufficient for the hand layup operation. The laminate was left for curing overnight before demoulding. The demoulding operation was performed easily, with the part leaving the mould without major effort or causing any damage. Figure 5.6 presents both the lamination process using the sheet metal mould and the finished part.

In the first evaluation by visual inspection, the fibreglass part succeeds to shape the model geometry, with a fair surface quality. No major issue is noticeable because of the use of a SPIF mould instead of a conventional mould, although some forming punch marks are still visible at the finished part. Some lost of contact points can be spotted near the ribs, although they relate to faults in the lamination operation and not to the mould itself.

Fig. 5.6 Hand layup mould during lamination and finished trimmed part

For dimensional control, the hand layup moulded parts were measured using a laser scanner and compared to the CAD model. The maximum deviation between the fibreglass part and the CAD model is +5.1 mm, with and average of +1.2 mm. When comparing to the real mould geometry, the maximum deviation is −2.6 mm.

In what regards the surface quality, some failures were found along the part, mainly on the higher slope reinforced wall. However, these faults are related to laminating issues and not to the use of the sheet metal mould. The faces moulded with tool contact have some minor marks from the sheet metal SPIF-related irregularities mainly on the lower slope wall. Notwithstanding these issues, the surface quality is generally considered fair along with all part.

The possibility of using stand-alone sheet metal parts for the development of open contact moulds was validated. Sheet metal mould is a reliable alternative to conventional tools for the manufacture of composite parts and compatible with typical accuracy levels. It is possible to achieve a reasonably good part quality even when laminating composites on unfinished SPIF sheet metal tools.

5.2.2 Compression Moulding

The compression moulding technologies are suitable for processing different materials. One of the most common usages is found in processing thermoset materials, fibre-based plastic composites and cellular materials. Different compression moulding techniques can be applied, using moulding materials in different states: bulk, sheet or granulate. In such a way, different techniques are suited to mould any geometry from thick, solid shapes to thin-walled shapes. Generally, the compression moulding process involves applying pressure to force the material contact with all mould surface areas, while heat and pressure are maintained until the material cures. Beyond the free-form moulding capability, a major advantage when compared to other manufacturing processes is the possibility of moulding variable thickness parts [4, 7, 8].

Fig. 5.7 Compression
moulding operation principle

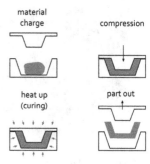

Compression moulding operation typically uses two-part moulds, defined by a cavity and a core. Figure 5.7 represents the basic operation of the moulding process. Depending on the used materials, the process may use heated moulds or a two-step process where the material is first compressed and then cured in an oven. The first approach is usually performed on hot press machines and the second approach is a regular press and an oven. Some moulding procedures, namely, when using thermoplastic matrices, preheat the material at a specific temperature to soften it, promoting and easier compression and mould filling. Throughout the process, heat and pressure are maintained until the polymer has cured. During the compression cycle, it is common to reopen the tool to let vapour escape. The mould may include venting features to better deal with this phenomenon [9, 10].

A relevant benefit of the compression moulding is that it discards relatively small waste, thus providing advantages when moulding with costly materials. Besides, compression moulding techniques are capable to mould extra large and complicated components. Finally, a major benefit of the compression moulding process, particularly when moulding foams or other cellular materials is the ability to control the material compression ratio and achieve a specific part density [10].

Typical compression moulding parts thickness varies from only 1 mm to over 25 mm, allowing shaping thicknesses variations along the part. The compression moulding part requires the use of a minimum draft angle from 3° and a minimum 2 mm radius in all edges. The process allows to free-form surfaces, including corrugated features, bosses or ribs [4].

Generally, the compression moulding loads vary from 0.1 to 10 MPa, increasing up to 15 MPa for some specific materials [4, 8]. This huge variation depends mainly on the compression moulding technology variant used and the moulding material. The loading range of the compression moulding process may limit the use of sheet metal tools, even when using thicker or stiffer materials. Still, ISF stand-alone tools may be suitable for some lower pressure operations and ISF rapid tooling with reinforcements may be widely applied.

5.2.2.1 Compression Moulding of Cork

A particular application of the compression moulding technology deals with the processing of cork composites. These materials commonly use 0.5–1.0 mm grain size granules, bounded by a polyurethane (PU)-based resin. The granulated cork density varies from 60 to 160 kg/m^3 and can be agglomerated to 140–600 kg/m^3 by compression moulding operations. Due to the absence of waste material, this manufacturing process allows achieving fairly complex geometries for medium to high quantities [11].

The compression moulding of cork may be performed in a two-step moulding operation, since the resin curing time is significant. Material mixtures used in compression moulding use cork resin ratios around 9:1. The material mixture adds water to the cork granules and binder for moisture cured polyurethane, enabling the aggregation of the powder and aiding compression and mould filling. Due to the great density difference between the material mixture and the finished part, the mould cavity must assure the volume for the bulk material. The compression is performed and the mould is locked in the final position to be heated for binder cure [12, 13].

The definition of a part geometry for an agglomerated cork piece follows the general guidelines for compression moulding. Special care must be taken regarding the minimum thickness to avoid fragile areas. In addition, minimum detail should also be sized according to the used granulate size.

For the evaluation of the SPIF cork compression moulding mould concept, a part was design based on the reference geometry. The material volume is reduced by designing the part as a thick shell. Along the smaller slope wall direction, the part uses a 26 mm thickness which increases to 37 mm in the lateral walls. Further, indentations features are included in the external surface of the part.

The designed tool for the cork compression moulding SPIF rapid tooling research uses a two-piece mould with a flat parting line. The mould is designed for a vertical operation, with the part being moulded upside down. One part of the mould is formed by a cavity with the complementary surface of the upper side of the designed part. The opposite part is formed by a core with the complementary geometry of the part hollow to mould the thick wall part. These two parts are the moulding essentials, designed as sheet metal containers. Since the raw material volume is much higher than the empty cavity, the mould assembly is completed with a frame box to allow the placing of all the uncompressed cork mixture. This frame box is defined to be manufactured by a flange-type SPIF part with a manually bent extension. For a more convenient operation, a support box is included in the assembly and a clamping system is added to lock the closed mould under pressure for resin cure.

Regarding the thermal behaviour, the resin curing deals with temperature below 200 °C and cure times from 2 to 12 h. These values are compatible with any mould thickness and, due to the long heating time, the heat-up rate is little significant.

From the mechanical point of view, in the cork compression moulding, moulding loads are typically kept in the lower range of the refereed interval. Depending on the intended part density, the moulding load may go up to 2.0 MPa [12]. Considering the compression of the cork alone, applications where the compression ratio is kept below 2, the differences in the compressive stress are not noticeable. In these cases, the moulding load is typically inferior to 0.6–0.7 MPa. In higher compressive ratios, densification occurs and the compressive stress increases abruptly. In compressive rations below 1.1, the moulding load is much lower. As the cork granulated mixture density is much lower than the cork itself due to the air between material granules, the moulding loads are even lower [13].

As the process pressure surpasses, the expected maximum loading capacity of the 2 mm blank aluminium sheet metal parts, special care must be taken to the tool design. In such a way, the support box was dimensioned so that the cavity sheet metal part is both supported around the perimeter and in the bottom. In what concerts the moulding load application, the core top allows a table-top support through a back plate. This operation principle ensures a better loading distribution. Still, mechanical simulation was performed to test the mould behaviour under compression. The moulding pressure was considered for a low reference value. This consideration was further supported by a moulding operation aiming for a low-density part. In such a way, the mechanical analysis considered a distributed load of 0.2 MPa. The test load considered the moulding compression itself. The mould locking was not considered since the force reduces after compression is stabilized in low to medium density parts.

Finite element method was used to test the mechanical behaviour of the tool. Meshing was done considering a solid mesh using tetrahedral elements with four integration points. The used element size was settled to 2 mm with a maximum deformation aspect ratio of 6. The mesh distribution was defined using curved based mesh, granting a minimum of eight elements in a circle and considering a maximum element size grow ratio of 1.6. The mesh was defined with around 300000 elements, with 93% of the elements with an aspect ratio under 3.

Fixtures were established to represent the moulding operation. The area in contact with the support box was fixed. In addition, the vertical downward movement of the cavity bottom and the vertical upward movement of the core top were restricted to simulate contact with the press plates.

Considering the 0.2 MPa loading and the described fixtures, the tool suffers a maximum stress of 180 MPa in the edges between the lower slope wall and the sidewalls and in the centre of the sidewalls. Although the value is larger than the yield stress, it only occurs in small areas. The lower value in comparison to the analytical calculation is due to the inclusion of rib indents in all walls which reduces the flat areas. The maximum displacement occurs in the centre of the sidewalls with a value of 0.6 mm.

The manufacturing process of the mould started with the two backing plate water jet cuts. One backing plate was used only for the core part of the tool and the other used both for the cavity side and the flanged box. The forming process of the three SPIF parts was done using a 12 mm spherical punch with a 0.5 mm vertical forming step in a helical tool path strategy. The feed rate was set to 2500 mm/min on the largest parts and to 2000 mm/min on the core side of the mould due to the smaller forming area. The container configuration parts were formed in a single-stage tool path strategy and the flanged configuration part was formed using a multistage strategy. For that, the SPIF operation was performed using a three-stage strategy. The first stage formed the sheet to a 45° wall, the second to a 70° wall and the firth to a vertical wall. The forming process considers the CAM preparation and NC programme compiling to the SPIF-A machine, machine set up with a tool change and backing plate change, sheet clamping, machine power up, sheet referral, forming operation and part release and cleaning. The process took just over an hour and a total energy consumption of 7.2 kW.h including close to 45 min of effective forming operation. After incremental forming, additional operations were performed to complete the mould assembly. A support box was built using four equal MDF pieces. Both the cavity part and the frame box were drilled to be fixed in the support using self-tapping screws. The frame box and the core part were also drilled in position for the mould locking mechanism. Complementary forming operations were performed to finish the frame box and the core side of the mould. The core part blank corners were cut, and the sides were bent using a manual sheet metal bender. The frame box height extension was manually bent around the frame box flange while riveting. The total fabrication time for the mould was 4 h 22 min being 56% for the SPIF process. The total energy consumption was 17.2 kW.h, and the total material cost was 20.00€, being 12.50€ for aluminium sheets. Figure 5.8 presents the complete tool assembly.

The mould sheet metal parts were measured by contact using a touch-trigger probe on a 3-axis CMM. Maximum deviation on the core side of the mould is +6.0 mm at the top edge of the longer higher slope wall. The average deviation is +1.0 mm. On the cavity side of the mould, the maximum deviation is +7.2 mm, with an average deviation of +1.1 mm. As in the hand layup mould, the measurement analysis shows accuracy improvement at the indent-reinforced walls. Besides, the addition of the reinforcement indents also benefits the wall flatness.

The mould operation was performed in two-stage operation. The compression was performed in a manual hydraulic press before locking the mould, and the cure was performed with the closed mould in an oven. A silicone-based release agent was used in both mould sides, along with tailored pieces of demoulding paper positioned in plannable areas.

Fig. 5.8 Compression moulding tool assembly and finished agglomerated cork part

Aiming for a $180\,kg/m^3$, the part weight is 167 g. Considering a 9:1 weight ratio, the cork composite mixture used 150 g of cork powder and 17 g of bounding resin. In order to add some moisture to the powder mixture, 10 g of water were added.

The cork mixture was added to the mould cavity with the frame box extension, filling it to approximately 30 mm height. The core sheet metal part was placed on top of the mould and pressed in a 15-tonne manual hydraulic press. During the compression operation, the maximum applied force reached close to 500 kg.f corresponding to a moulding pressure of 0.15 MPa when considering the 180×180 mm projected area. No visible deformation was noticeable during the moulding operation. During the mould compression, a small amount of material was lost by the side locking holes. Yet, the amount can be despised when compared with the moulding material volume. The mould was locked while under pressure using the side screws. The frame box supports the closed mould without core part slide during the press force release. The closed mould was removed from the press and cured on an oven for 2 h with a temperature set for 140 °C with maximum peak of 150 °C for 4 h.

The moulded part released with ease from the mould due to the use of the demoulding paper. Nevertheless, some wrinkles are formed in the paper during the compression moulding than harm the part surface. However, the use of the silicone demoulding agent alone proved to be insufficient for the cork compression moulding operation with the compression ratio of 1:2, causing damage at the part surface during demoulding operation. Figure 5.8 presents the demoulded part after cleaning the unagglomerated cork powder.

After compression, the mould parts were remeasured to check for permanent deformation. The core side of the mould suffered a permanent deformation close to 3.5 mm at the centre of the edge farther from the core feature. Besides, a mark from the pressure distribution plate of the press is noticeable at the part. The cavity side was deformed by 4.0 mm at the centre of the lower slope wall. All remaining walls suffer no permanent deformation.

For the part evaluation, the first visual inspection was performed. The evaluation assesses the mould filling, part apparent density and general part quality. The part succeeds to define the overall geometry definition, with only minor failures at the sharp edges, small defects due to wrinkles in the demoulding paper at the bottom side of the part and some high concentration binder spots noticeable at surface. Nevertheless, the surface quality is good and cork granulated congregate apparent distribution is uniform.

The part was weight for density evaluation. The total part weight was 166 g, resulting in an effective global density of $180\,kg/m^3$. During the moulding process, the material lost was negligible. A manual compression of the part surface suggests a uniform material behaviour, and so a uniform material density. The part was measured for accuracy evaluation. Despite some significant deviations, the cork part was moulded with an average deviation of −0.4 mm. However, the largest deviation along the part surface goes up to 6.3 mm at the centre of the lower slope wall, due to the mould deformation. The biggest deviation is found in the lower edges with the real part −13 mm smaller than the CAD model. This large value is partly due to the failure in defining the part bottom sharp edge.

> The use of sheet metal moulds for cork compression moulding operation was validated. Some relevant issues were found at the moulded parts. Nevertheless, some of these problems, particularly the binder spots and the wrinkle marks, result from the compression moulding operation and not to the mould itself. Only the miss definition of the part edge is affected by the mould, due to the considerable gap between the frame box and the core side. Even so, the use of sheet metal moulds is an attractive solution for compression moulding with cork or other powder and binder-based composites, mainly when the production volume requires low-cost tooling.

The mould fabrication has reduced time and cost and has the potential to achieve relatively complex parts. The inclusion of sheet metal parts with other parts increases the geometry limits and, in specific situations, lead to a faster mould development and lower tooling cost.

From the mechanical point of view, the 2 mm sheet metal alone was sufficiently rigid although not tough enough to support the moulding loads, leading to permanent deformation. On the core side of the mould, the permanent deformation results from an uneven pressure distribution. The addition of a correct size pressure plate could eliminate this issue. On the cavity side, the deformation is due to the geometry itself. The mechanical behaviour could be improved by either use a thicker sheet or by adding a filling support using sand or other porous mixtures.

Apart from the defects that result from the mould deformation and oversized gap between the core and the frame box and operation issues, the part quality is good. The unaffected surfaces have an accuracy of ± 2 mm and a good surface finishing. The punch marks on the sheet metal surface are not visible on the moulded part.

5.3 Low-Pressure Polymer Processing

The injection moulding, blow moulding and extrusion are the polymer processes more frequently used in polymer processing. However, there are other plastic processes particularly important in manufacturing industries such as thermoforming and rotational moulding. These processes are typically characterized by low-pressure operation and use moderate to high temperature, depending on raw material. In such a way, ISF rapid tooling techniques are feasible for the development of moulds for these processes.

The strength and stiffness of the selected material is compatible with the typical moulding loads. Besides, the low thermal inertia of the sheet leads to fast and cost-effective cycle times, supporting the use of ISF tools.

5.3.1   Thermoforming

The thermoforming process allows the production of plastic parts by shaping a polymer sheet or film to a mould. The polymer is heated to a pliable state and then pushed to the mould surface. Almost any thermoplastic can be thermoformed, where the most common is the use of high impact polystyrene (HIPS), acrylonitrile butadiene styrene (ABS), high-density polyethylene (HDPE) and polymethyl methacrylate (PMMA), commonly known as acrylic. Different methods exist, being vacuum forming one of the most used to the production of small series of 3D-shaped plastic parts. Despite using a much simpler mould than other polymer processing technologies as injection, mould cost and development time still have a considerable weight in the development of a product. Still, the thermoforming processes can be used for a high- or low-volume production and it is commonly used for prototypes production [9, 14].

Figure 5.9 represents the basic operation concept of a vacuum forming process. A flat thermoplastic sheet is fixed on the forming machine granting a peripheral air sealing. The plastic sheet is heated either before or after the clamping, depending on the machine used. After being heated, the plastic sheet is stretched by a blowing operation before the mould rise, mainly when forming with core moulds. Forming with cavity moulds can dismiss the pre-stretching. Vacuum is used to shape the material against the mould before being cooled by an air blower. After cooling, air is blown to help demoulding and the mould descent before part release. The moulding process is relatively slow, with cycle times up to 5 min. One of the disadvantages of thermoforming is the part wall uniformity. As the moulding step involves typically only one-part tools, there are more difficulties controlling the wall thickness.

Thermoforming is a low-pressure process. Some thermoforming techniques may use additional pressure up to 0.3 MPa [9]. In the case of vacuum forming, the maximum moulding force occurs at full vacuum. Accordingly, mould pressure is limited to a reference value of 0.1 MPa.

Fig. 5.9 Thermoforming operation principle

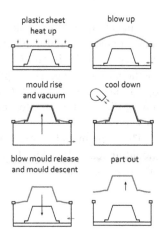

Most thermoforming tools are single surfaced. One surface of the plastic sheet is forced against the mould and the other surface remains unimpeded. Both male and female moulds can be used in vacuum thermoforming operations. Both mould types have complementary geometries of the plastic part, controlling either inside or outside surface. Moulds are usually made by casting aluminium or most frequently milling aluminium, high-density rigid polyurethane foams and wood. Sprayed metal, electroformed nickel, hydrocal and other materials and technologies are also used to make thermoforming moulds [15]. These processes and materials make most mould designs expensive and they take a long time to be built. Besides, high thermal inertia moulds can spoil the continuous operation if cooling is not performed.

The female or negative mould uses a cavity with the complementary geometry of the outside of the plastic part. It is used when the outside of the plastic part is to be controlled. The male or positive mould uses a boss with the complementary geometry of the inside of the plastic part. It is used when the inside of plastic part is to be controlled. Both mould configurations must be drilled in the lower points to allow the vacuum to pull the plastic sheet. Highly complex parts can include both positive and negative features in the same mould. The opposite side of the mould tolerance in inferior as thinning occurs during thermoforming. Besides, due to the strain during the forming operation, textured plastic sheets are highly affected by thermoforming operation, mainly when forming tall parts.

As vacuum thermoforming pressure is low, it is possible to replace conventional tooling materials and designs by sheet metal-based tools. ISF can be used to shape sheet metal to the desired surface enabling the design of sheet metal moulds. Apart from being a potentially faster and more economical tooling process, a low thermal inertia may benefit the mould operation. Nevertheless, the fabrication of the sheet metal mould as some geometry-related issues.

Thermoforming parts are shaped like open shell structures from small as a few millimetres to a few metres. Most parts use thickness up to 3 mm, while thicker sheets up to 5–6 mm can be used. A minimum 5° draft angle should be used in order to be possible to form parts with any material, namely, crystalline polymers, mainly when forming with positive moulds. As the mould may have texture due to technological reasons, 1° additional draft should be added for every 5 μm in texture depth. Due to material compression during cooling, parts from negative moulds could have smaller or even zero draft. Lastly, one can say that the larger the draft angle, the better for the finished part [15].

The moulding process allows to shape the plastic sheet to a free-form surface, with limitation of obtaining sharp edges. In what concerns fillet radius, to minimize corner stress concentration on the formed part, internal corner radii should have a recommended minimum of 80% of the plastic thickness. Like in draft, the bigger the fillets on both side and top faces are, the better for the finished part. Fillet radius can also be influenced by technological limitations on the mould-making process.

Lastly, a draw ratio of the part geometry also influences its feasibility, where the draw ratio is determined by the surface area of the part over its footprint. Generally, every feature on a part should avoid being narrower than it is tall. Besides, as the draw ratio gets larger the radii will almost always have to be increased.

For the thermoforming sheet metal mould development test, a part was designed based on the reference geometry simply by adding fillets to the edges. The internal radius used between side faces was 15 mm and between side and top faces was 6 mm. The part was set to be thermoformed using a 3 mm plastic sheet. For the evaluation of the use of stand-alone sheet metal moulds for thermoforming operations two approaches were followed, operating with a positive and with a negative mould. The negative mould was designed straightforwardly from the part geometry. The positive mould was designed considering the thinning predicted by the sine law (Eq. 2.1).

Regarding the mechanical behaviour of the tool has the moulding pressure near the maximum feasible pressure, numerical simulation was carried out to forecast the stress and mould displacement. The simulation of the mould mechanical behaviour considers the worst-case scenario for a vacuum forming operation, where the full moulding surface is under the maximum plausible pressure. Thus, a uniform pressure of 0.1 MPa on material contact surfaces is considered for the mechanical analysis. The simulation considered not only the sheet metal parts but also the MDF support box with a global contact between parts defined with no penetration and a static frictional coefficient of 0.2. Meshing was done considering a solid mesh using tetrahedral elements with four integration points, similar to the one used on the compression moulding test. The total mesh has just under 900000 elements where 98.2% have an aspect ratio below 3 on the positive mould and 1650000 elements where 98.3% have an aspect ratio below 3 on the negative mould.

The maximum displacement determined is 2.53 mm on the positive mould and 2.27 mm on the negative mould, consistent with the forecast by the analytical calculation. The stress results present a maximum value of 180 MPa at the part boundary and close to 130 MPa both at the centre of the flat areas and at the side fillets. Although the calculated displacement is significant, it is lower than the SPIF typical accuracy. Thus, increasing the mould thickness should not benefit the thermoformed part accuracy as despite being more rigid, may be less accurate. With regard to stress, although the value is close to yield, it only occurs on specific points and small areas. Furthermore, calculus considered a pressure of 0.1 MPa (pure vacuum), situation that is never fully achieved in thermoforming process.

Alike the previously presented tests, the mould manufacturing process started with the backing plate water jet cut operation. For each mould configuration, the sheet metal part was formed using a helical SPIF strategy with a 12 mm punch running at 3000 mm/min with 0.5 mm vertical increments. The sheet metal parts were drilled to allow vacuum during thermoforming when the plastic sheet seals to the mould surface, boring 81 mm holes on the negative mould and 31.5 mm holes on the positive mould. After the sheet metal work, support boxes were cut and assembled from MDF boards. The positive mould support consists of a simple board cut with a draft angle. The negative mould support was assembled from two sets of two boards, mounted with a draft angle to facilitate the thermoforming operation. Total fabrication time for the positive mould was 1 h 27 min being only 58% for SPIF process. The total energy consumption was 6.6 kW.h, although a large proportion is derived from the water jet cutting. The total material cost was just 8.00€, being 2.50€ for the backing plate, 4.50€ for the aluminium sheet and 1.00€ for the MDF. The negative mould fabrication time was 1 h 55 min with and energy consumption of 6.7 kW.h and total cost of 8.50€. The difference in manufacturing time, energy consumption and cost result from the additional carpentry time during the support box manufacture.

After manufacture, SPIF sheet metal moulds were measured by contact using a touch-trigger probe on a 3-axis CMM. The maximum deviation is +2.9 mm on the positive mould and +6.8 mm on the negative mould. Average deviation is +0.3 mm on the positive mould and +1.8 mm on the negative mould. Furthermore, both moulds were remeasured after being used to check for permanent deformations caused due to vacuum. No major variations are found with average of 0.05 mm between the first measurement and the one done after use.

For a comparative analysis, conventional milled moulds were manufactured using high-density polyurethane blocks. On average, the SPIF moulds took 70% less time to produce than the conventional ones. In what concerns material cost, the SPIF manufacture allowed a reduction of almost 80%. In terms of energy consumption, the SPIF process allowed a reduction of 20%. For mould validation, various tests were performed using both the SPIF sheet metal moulds and the conventional moulds. The moulding surface was sprayed up with silicone-based demoulding agent and a vacuum forming machine was used. Figure 5.10 presents one of the thermoforming tests performed with the positive sheet metal mould, thermoforming with 3 mm clear PMMA plastic sheet. The forming operation works without noticeable differences between the sheet metal and the conventional tools. The plastic sheet forms properly to the moulding surface and the mould suffers no visible deformation during the vacuum. The vacuum holes allow to pull the material against the mould and no issue is caused by the material contact. When forming a series of parts, minor heat increase occurs in the sheet metal moulds, resulting in a faster cooling rate of the plastic parts. After forming, the parts release with ease from the moulds. Besides, in the sheet metal moulds, no harm is enforced neither to the moulds or the parts when no demoulding agent is used.

After moulding operation, the part is finished. Finishing operation includes the release agent cleaning and trimming. The trimming operation is performed using a bent saw, and the cut is manually deburr. Figure 5.10 shows one black PMMA part after finishing operations.

The plastic parts formed in both the sheet metal moulds and the conventional moulds were first evaluated through visual inspection. All the four moulds succeed to shape the plastic to the desired geometry. Nevertheless, small differences are noticeable both between the positive and the negative moulds and between the sheet metal and the conventional moulds. The major differences between the sheet metal mould and the conventional deal with the surface flatness and finishing. The tent effect lack of accuracy in the sheet metal parts has a slight negative influence on the parts' aesthetics. In what regards the surface, both the punch skinning marks from the sheet metal forming operation and the milling increments from the conventional manufacture are visible in the finished part, without major differences.

Fig. 5.10 Thermoforming process with clear PMMA using the sheet metal positive mould and trimmed black PMMA part thermoformed on the positive mould

For dimensional control, the thermoformed parts were measured by contact on a CMM. On each part, the surface in contact with the mould is measured. On the part formed on the positive mould, maximum deviation between formed part and CAD model is 14.7 mm with an average of 0.7 mm. On the part formed on the negative mould, maximum deviation is 11.4 mm with an average of 1.8 mm. When comparing parts to the real mould geometry, maximum deviation is 5.8 mm on the positive and 10.6 mm on the negative with respective average of 0.1 and 0.2 mm.

For a mechanical behaviour evaluation, the parts formed with the two tools concept were tested. For that, specimens were cut from parts formed on the positive tools with 2 mm clear PMMA. Four specimens were cut from each part, three from the lower slope wall and one from the top wall. A tensile test was used to compare the two parts. No differences were found between the mechanical behaviours of the two parts. All strain–stress curves follow the same profile with similar ultimate strength value. However, the specimens cut from the part formed with the conventional mould tend to break sooner due to bigger cut defects.

The proof of concept tests validates the use of the SPIF process as a rapid tooling process for thermoforming operations. Sheet metal moulds made by SPIF are a reliable alternative to the conventional moulds. Despite some significant deviations still occur at some points (mostly peripheral), general appearance of plastic parts is reasonably good and average dimension is accurate.

The thermoforming moulds have low complexity in terms of design and are mostly compatible with ISF design restrictions. The use of sheet metal tools reduces both the bulk materials used as the waste material. Tooling times are also significantly reduced comparatively to the moulds produced by conventional milling technologies. Besides, the test reports the potential of alternative moulds for thermoforming offering a better use of material and better energy efficiency. Finally, the faster thermal cycling is the greater advantage of sheet metal moulds.

Both the mechanical and thermal behaviours of the sheet metal moulds are well suited for the thermoforming operation. Regarding the thermal performance, the low thermal inertia due to the little volume of material may have advantages in the temperature cycles, mainly in the cooling process of the plastic. From the mechanical point of view, stand-alone sheet metal moulds using medium to thick sheets can support the plastic forming forces on small- and medium-size moulds.

When comparing the sheet metal moulds manufacturing process against conventional moulds, the new approach is more economical in both development time, energy consumption and material cost.

Besides, moulds total manufacture time and material cost are low and yet largely influenced by some parallel operations. As accessories like backing plate for SPIF operation and MDF support boxes can be used for more than one mould with similar projected area, the use of sheet metal moulds assumes even a more interesting panorama. Relatively to energy, a slight reduction was archived, although not as significant as the time and material savings for the work done.

On the other hand, despite being acceptable, the thermoformed parts' quality is superior in the conventional moulds, mainly in what concerns geometric and dimensional tolerances. However, the major differences between the operation with the different mould concepts may be reduced by finishing operations on the sheet metal moulds allowing the improvement of the part accuracy. From the mechanical point of view, no differences are found between the parts formed in sheet metal moulds and in conventional tools.

In such a way, the SPIF rapid tooling applications for thermoforming operations are promising, mainly for the development of prototype tools or small series production [16].

5.3.2 Rotomoulding

The rotational moulding processes allow the production of hollow plastic one-piece parts. Two different approaches can be used in rotational moulding. On rotocasting, a thermoset material is used. On rotomoulding, a thermoplastic polymer powder is charged in a mould and then heated while the mould rotates axially or biaxially inside a forced convection oven. No pressure is used apart from a low contact pressure during the rotation of the heated melt. After the heat-up cycle, the mould is cooled down at room temperature or under a water spray while rotation continues before being demoulded. The moulding process leads to stress-free parts despite the poor dimensional tolerance control and low mechanical properties. Figure 5.11 illustrates the rotomoulding operation principle. Typical process cycle takes between 10 and 20 min for the heat-up and 10–20 min for the cooldown. The total process time adds the material loading and the part removal. The most common materials use is low-density polyethylene (LDPE) in powder size mesh, with a melting point around 100 °C. The average wall thickness is determined by the amount of material in the mould. The most common rotomoulding machines are the batch and the carousel

Fig. 5.11 Rotomoulding
operation principle

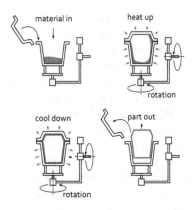

types. The carousel type use multiple stations for material loading, heating, cooling
and product removal. These machines are usually automated and used for larger
productions. The batch type is manually operated to produce a part at a time and is
typically used for a low production. During the rotomoulding process, the rotation
speed is usually kept constant with typical values between 3 and 15 rpm. The ratio
between the major axis and the minor axis varies between 8:1 and 1:5, according to
the part configuration and orientation [9, 17, 18].

The rotomoulding temperature cycle is a strong influence on the moulded part.
Apart from influencing the moulding time, the mould design should ensure a uniform temperature distribution. During the heat-up period, the inner surface is slightly
cooler than the outer surface and the air temperature inside the mould is even cooler.
Until a certain period, the material inside the mould has not reached the tacky temperature and the speed rotations are not critical. When the inner air temperature reaches
the material melting point, the heat-up ratio plateau occurs as the material absorbs
most of the heat. During this period, the material starts to adhere to the mould so the
rotation speed and speed ratio are critical to allow a uniform thickness distribution.
After all material has adhered to the mould wall and the melt is almost complete, the
plastic starts to coalesce and densify. The coalescence process is completed to a uniform melt before removing the mould assembly from the oven. During this period,
air bubbles are trap in the plastic. The cooking time should be enough to allow the air
to exit but must prevent over-cooking as degradation starts rapidly once the bubbles
disappear. After removing the mould from the oven, the cooling process begins. The
air temperature drops quickly until the material starts to solidify. The cooling rate
affects the material properties. A second temperature plateau occurs when the plastic
solidifies and crystallizes, giving off heat. A final stage in the cooling occurs after
complete solidification when the plastic shrinks and creates an isolating layer of air
between the mould and the material. When the material reaches a temperature close
to the ambient, the part can be removed from the mould [17, 18].

The moulds used for rotomoulding are usually thin shell-like metal structures
manufactured by electroforming nickel, casting aluminium, aluminium sheet or

steel sheet. Electroformed mould is mainly used for small detailed parts. Casting aluminium is used for medium-size moulds. Sheet metal moulds are also commonly used in large moulds, particularly steel due to the ease of welding. The mould assembly may include the moulding surface to be mounted on a mould frame to help its assembly and clamping. Two-part moulds are mainly used, defining a hollow surface with the straightest parting line possible. More complex assemblies are possible to allow more intricate geometry parts. The mould assembly must include a clamping system, as well as closing guiding features. The mould typically uses a venting channel to allow air to exit the inside during the heating process and reenter during cool off, minimizing the risk of creating blowholes in the part or distorting the mould. Different mould surfaces including texture can be used, while polished surfaces should be avoided. For an effective moulding, a release agent must be used [9, 17]. The lower tooling costs when compared to other high-pressure processes validate the possibility of production of few pieces of large dimensions or complex parts that are otherwise not economically feasible. Still, the tooling cost and development time still have a strong role in the process practicability.

The great advantage of rotational moulding is the possibility of produce one piece with hollow parts of complex shapes in a flexible range of dimensions. Thickness varies from 1.5 to 5 mm and the part is detailed only in the outer side. Draft angle is recommended to facilitate the part removal; however, parts can be manufactured with only a small draft angle or even slight undercuts as the material shrinks away from the cavity mould surface during cooling. Intersecting planes should avoid sharp edges, with recommended radii equal to the thickness. Parting lines should be as straight as possible and large flat areas should be avoided. The high-stress areas of the parts should be kept away from the parting line and the part can be reinforced using corrugated surfaces, stand-up bosses or kiss-off structures. These geometric features require the use of a draft as the material shrinks to the mould in the male geometric features of the mould during cooling [9, 13, 17].

The rotational moulding part was designed by closing the reference geometry to a six-side box. Since flat areas should be avoided, a quarter circle arc-shaped slot 20-mm-deep deboss with a 20° draft angle was enclosed in the bottom side. The upper side of the part was also changed from a simple flat configuration to create an opened kiss-off with a 22 mm by 30 mm area, 25 mm distant from the bottom side. This linkage between two-part sides is common in rotomoulded parts to serve as a structural reinforcement and can lead to only a contact point or an open window.

The mould was designed as a two-part mould fabricated in aluminium sheet metal by SPIF. The sheet metal parts were modelled according to the sine law allowing the simulation of both the mechanical and thermal behaviours of the tool. From the mechanical point of view, the tool suffers a maximum displacement of 2.3 mm and a maximum stress value of 180 MPa when under a 1 atm hydrostatic pressure. Considering the thermal simulation with a 2 mm polyethylene plastic layer, the heat-up temperature in a 250 °C oven is 12 min, with the plastic material reaching values from 184 to 222 °C and the sheet mould reaching an average temperature of 209 °C.

The manufacturing process includes the cut of individual backing plates for each mould side. The blanks were cut, including the trimming of the corners to best fit the rotational moulding system work volume. For the SPIF operation, a single-stage helical tool path with a vertical constant increment of 0.5 mm and a feed rate of 3000 mm/min using a 15 mm ball tip tool was used. To improve the part accuracy of the kiss-off area, a local support was also used, being manually positioned under the blank at the moment the punch reaches a depth of −20 mm. Additionally, an over forming procedure was used to improve the accuracy of the kiss-off area. After forming operation, the mould sheet parts were drilled to allow both the positioning and closing and clamping. The mould was finished by filling up the undesired small radius created at the blank edge during the forming operation using cold welding mass. The fabrication time for the mould was 2 h 26 min, being only 54% for the SPIF process. The required time to use the mould adds the cold welding mass curing time. The total energy consumption was 11.0 kW.h, although a large proportion is derived from the water jet cutting. The total material cost was just 20.00€, being 5.00€ for the backing plates, 7.50€ for the aluminium sheet, 5.00€ for the cold welding mass and 2.50€ in fasteners.

After manufacture, the SPIF sheet moulds were measured by contact. The maximum deviation is +2.9 mm on the lower side and +6.8 mm on the upper side of the mould. Average deviation is +0.3 mm on the lower side and +1.8 mm on the upper mould side. After measured, the mould was closed and heated up to 250 °C to check for stiffness issues due to the internal air heating or thermal treatment distortions. No failure occurs during the heating and cooling process. The parts were remeasured without registering dimensional changes. In such a way, the manufactured mould was validated and approved for operation tests.

The mould operation tests have been developed for both a rotocasting process using a polyurethane resin and for a rotomoulding process using a low-density polyethylene powder. The tests were performed using a rotational moulding system between 4 and 10 r.p.m. with perpendicular rotation axis and a speed ratio of 1:6. The rotocasting operation was performed at room temperature under continuous rotation during the 30 min of pot life and regularly rotated by 180° during the additional 3 h of gel time. The rotomoulding operation was performed in a forced convection oven. Figure 5.12 presents the open mould filled with LDPE powder 35 mesh inside the rotomoulding system. The mould was rotated for 12 min in the oven with temperature between 210 and 250 °C and cooled down at room temperature for 15 additional minutes.

The PU part geometry reproduces in detail every design element. An over thickness is noticeable at the part thinner edge due to the material run during the gel time. Despite the low resin viscosity, no leakage occurs during the moulding operation. However, a concave structure with small burrs is noticeable at the parting line. The LDPE part, presented in Fig. 5.12, is finished with higher quality. The overall material distribution along the moulding surface is faultless, the surface finishing is acceptable and the overall geometry is moulded as expected. No significant burrs are found along the parting line although a small undensified volume occurs due to the cold spots caused by the filling mass. The open kiss-off at the part centre has better results than the ones from the rotocasting operation, with an open structure despite the miss defined boundary.

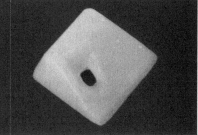

Fig. 5.12 Open sheet metal mould filled up with LDPE in the rotational moulding device and finished part after deburring

After moulding operation, the parts were finished. Finishing operation includes the release agent cleaning, deburring and improving external fillets. Deburring operation was done using manual tools, improving the fillets along the parting line, including the open kiss-off. Along the operation, major volume of undensified material was removed.

For dimensional control, the rotational moulded parts were measured using a hand laser scanner and compared to the CAD model. The LDPE part is on average 2.8 mm smaller than the CAD model, with deviation varying from −9.7 to +5.5 mm. The part under dimension occurs both due to mould inaccuracy and material shrinkage. When comparing to the real mould geometry, parts are on average 1.3 mm smaller than the mould, being only bigger in the slot features. The average accuracy is proper of typical rotomoulding process although some deviations, mainly on the bottom side are slightly greater than desired. This deviation occurs due to the impossibility of improving top radius of the lower mould size and due to a slight inaccuracy at the moulding surface contact to create the kiss-off structure.

For a thickness evaluation, parts are cut by a quarter. The LDPE parts present an average thickness along the cut of 2.3 mm, with wall thickness varying from 1.7 and 3.0 mm. The average thickness of the base plane and the lower slope wall is 2.1 mm. The higher slope walls average thickness is 2.7 mm. The higher thickness on the part is found at the corners between the base and the higher slope wall, with a maximum value of 7.5 mm. It is noticeable that the unidentified material is only found at the outside surface and the corner has a 4.6 mm solid thickness, minimizing the undesired effects of the cold spot. The edge between the base and the lower slope wall features the worst thickness distribution, with thickness down to 0.8 mm and failing to close the surface at some points. This matter occurs mostly because of material distribution issues due to the part low angle.

The proof of concept tests validates the use of SPIF as a rapid tooling process for both rotocasting and rotomoulding operations. Sheet metal moulds made by ISF are a reliable alternative to the conventional moulds. The sheet metal mechanic and thermal behaviours are suitable for rotomoulding operations, being confirmed both numerically and experimentally. Despite some significant deviations due to the inaccuracy of the forming process, general appearance of plastic parts is reasonably good and average dimension is accurate.

The mould total manufacture time and material cost are low and yet largely influenced by some parallel operations. As some tools like backing plate for SPIF operation can be used for more than one mould with similar projected area, the use of sheet metal moulds assumes even a more interesting panorama. The performed test uses two different backing plates while in parts with a more typical parting line could share the backing plate for both mould parts.

If comparing the mould manufacture with typical process, one can predict the SPIF sheet metal approach is both more economical in an energetic and material point of view, as well as in the development time. However, the mould accuracy is lower and leads to the worst surface quality.

Yet, the process accuracy and surface finishing can be improved. The accuracy, and surface quality as well, can be improved using multistage or other forming techniques, although they lead to a longer forming time. A fair compromise must be pursued. Besides, complementary operations can also improve the mould quality. Sanding operation alone can improve the surface quality. A full coverage of the moulding surface with epoxy mass associated to sanding and milling operations could lead to additional manufacturing time and cost but grant great results, although attention must be paid to avoid the creation of cold spots.

This new approach for the manufacture of sheet metal rotational moulding moulds allows to increase the possible geometric complexity. The developed case study used a mould fully developed by SPIF for proof of concept. Nevertheless, the incremental sheet forming technologies can be mixed with other processes, achieving great design possibilities.

The uneven thickness distribution on the mould walls due to the incremental forming process lead to a small uneven thickness along the moulded part due to faster heating times. Nevertheless, those differences do not unfeasible the rotomoulded parts and are minor than the ones caused by material distribution [19].

References

1. ASTM, Standard test method for tensile properties of plastics (West Conshohocken, PA, 2010)
2. ISO 10002-1 (2009) Metallic materials—tensile testing part1: method of test at ambient temperature

3. D. Cripps, T. Searle, J. Summerscales, *Comprehensive Composite Materials* (Elsevier, Open Mold Techniques for Thermoset Composites, 2000)
4. Molded Fiber Glass Companies (2016) Technical Design Guide for FRP Composite Products and Parts—Techniques & Technologies for Cost Effectiveness
5. C. Santulli, Alternatives for a hand lay-up composite structure: E-glass/epoxy adhesive joint or tapered laminate. J. Mater. Sci. Lett. **21**(24), 1959–1963 (2002)
6. M. Elkington, D. Bloom, C. Ward, A. Chatzimichali, K. Potter, Hand layup: understanding the manual process. Adv. Manufact. Polym. Compos. Sci. **1**(3), 138–151 (2015)
7. C. Lefteri, *Making It: Manufacturing Techniques for Product Design* (Laurence King, 2012)
8. Victrex (2016) Compression Moulding Processing Guide
9. D. Rosato, D. Rosato, M. Rosato, *Plastic Product Material & Process Selection Handbook* (Elsevier, 2004)
10. A. Shamsuri, Compression moulding technique for manufacturing biocomposite products, International Journal of Applied Science and Technol **5**(3), 23–26 (2015)
11. Amorim Cork Composites (2017) *Cork Solutions & Manufacturing Processes*
12. E. Fernandes, V. Silva, J. Chagas, R. Reis, Cork-polymer composite (CPC) materials and processes to obtain the same, WO Patent App. PCT/PT2008/000,051 (2009)
13. S. Silva, M. Sabino, E. Fernandes, V. Correlo, L. Boesel, R. Reis, Cork: properties, capabilities and applications. Int. Mater. Rev. **50**(6), 345–365 (2005)
14. J. Throne, *Understanding Thermoforming* (Hanser, Dunedin, 1999)
15. J. Throne, *Thermoforming 101* (SPE Thermoforming Division, 2006)
16. D. Afonso, R. Alves de Sousa, R. Torcato, Testing single point incremental forming molds for thermoforming operations, in *ESAFORM 2016* (2016)
17. R. Crawford, M. Kearns, *Practical Guide to Rotational Moulding* (Rapra Technology Limited, 2003)
18. LyondellBasell (2016) A Guide to Rotational Molding, 5717E/0715
19. D. Afonso, R. Alves de Sousa, R. Torcato, Testing single point incremental forming molds for rotomoulding operations, in *ESAFORM 2017* (2017)

Printed in the United States
By Bookmasters